BUILDING, MARKETING & OPERATING

a

PROFITABLE TAX PRACTICE

Third, Expanded, Edition

A practical, step-by-step working guide to a successful, efficient and lucrative professional tax practice

NATIONAL TAX TRAINING SCHOOL

Publications Division

Serving the Tax Profession since 1952

http://www.nattax.com

"This publication is designed to provide accurate and authoritative information in regard to the subject matter covered. It is sold with the understanding that the publisher is not engaged in rendering legal, accounting or other professional service. If legal advice or other expert assistance is required, the services of a competent professional person should be sought"

From a Declaration of Principles jointly adopted by a Committee of the American Bar Association and a committee of Publishers and Associations.

Published by:
National Tax Training School
Publications Division
PO Box 382
Monsey, New York 10952
Voice 1-800-914-8138 Fax 1-914-352-8138

Website: http://www.nattax.com
E-mail: info@nattax.com

ISBN: 0943790–69–7

CHAPTER HEADINGS

See Detailed Table of Contents at End of Book

A Word To The Reader

The original version of this work, BUILDING & OPERATING A PROFITABLE TAX PRACTICE, was first published in 1977 and reprinted several times after. A subsequent second edition significantly expanded the scope and contents of this work.

The present third edition has been completely revised and reorganized, a substantial amount of new material added and both the title and focus modified to reflect the enormous changes in the accounting and tax professions that have taken place in recent years.

Initially, the book was designed to serve as a hands-on blueprint for new tax practitioners—mainly graduates of National Tax Training School's basic and advanced tax training programs. But, to our surprise, the feedback we received indicated that even seasoned professionals with years of experience found much useful and valuable information and a wealth of practical tips in its pages.

Accordingly, in the second edition—and especially in this brand new revision—we took great pains to address the interests and needs of those readers as well. Whereas, originally, the primary goal was to guide the newcomer through the various steps of setting up a viable tax practice, we now focus extensively on providing proven ideas and know-how for growing an existing practice in both clientele and revenues.

If you are a new entrant into the tax field—or are presently working for an accounting or tax firm and contemplate going into business for yourself—you need not feel shortchanged. All the information, tips and guidelines that have helped so many of your predecessors build flourishing professional practices are still there, with a host of new ones added.. But this Manual will also show you how to take your business much further once you're firmly established. After all, isn't that what you're aiming at?

If you're an experienced professional with a going tax service business you may want to skip (or just skim) those portions of the first few chapters that deal with setting up a practice. But the balance of the book, comprising the overwhelming majority of topics covered, is addressed to both novices and old-timers alike. You'll find a goldmine of down-to-earth advice, proven strategies and tested marketing methods—all designed to boost your bottom line.

To get the most out of this book, don't just read through the pages. Use it as a hands-on working manual. Make marginal notes, underline or highlight business- and profit-building ideas and techniques you wish to implement immediately, or at a later date. Most important, try to formulate a clear, specific, plan of action to advance your practice from where it is now to where you want it to be. You can be assured that your efforts will be well rewarded.

Have a happy, productive, journey.

Acknowledgments

This work is dedicated to the thousands of NTTS students and graduates throughout the United States (and abroad) whose enthusiasm, professionalism and success stories have been a constant source of encouragement and inspiration to us for over four decades.

We are particularly grateful to the scores of NATIONAL TAX subscribers and customers—the men and women in the trenches—both students and nonstudents, who have supplied so much of the material in this book.

They graciously shared with us their experiences, patiently answered detailed surveys and questionnaires regarding fees, marketing methods, promotion strategies, hardware and software, etc. Many went out of their way to submit innovative ideas, point out potential pitfalls or volunteer other valuable 'inside' information...for no other reason than to extend a helping hand to their fellow professionals.

We are proud, indeed, to be part of this fraternity.

Chapter 1
INTRODUCTION

Three men were performing a similar task when a passerby stopped to observe them. After a few minutes he asked one worker, "What are you doing?" And the man answered, "I'm laying bricks." He then asked the second worker, "And what are you doing?" And the man said, "Why—I'm building a wall." Then he asked the third worker the same question, and the man looked up and replied: "I'm building a house of worship!"

It all depends on how you look at things: a lot of people may be performing a similar task, and one can think of it as menial, while the other sees it as inspiring—and the difference depends upon their outlook on life.

Author Unknown

The journey through life can be a pleasant and rewarding experience if we have the right attitude. A good attitude is like cork—it can hold you up. A poor attitude is like lead—it can sink you.

L. Kenneth Wright

General Introduction

Our system of taxation in America has a long and curious history; and in many ways the evolving tax system even mirrors the development of our country as a whole.

The early colonists rebelled against what they considered to be unfair tax legislation imposed by the British magistrates: a sugar tax in 1764, a stamp tax in 1766, and a tax on tea in 1773. All of these events, of course, led up to the Revolutionary War, and finally, American independence in 1776.

It takes time for a country to grow, and it wasn't until 1861 that our first federal tax law was signed, and that was to help finance the Civil War. With that as a base, income tax laws progressed, and with the advent of the First World War, taxes and tax legislation increased dramatically. This trend naturally continued through World War II and thereafter.

As 20th Century life in America became more complex; so, too, did tax legislation. The result, at present, is a body of law so intricate that few taxpayers can obey it without some expert assistance. Consequently, a great many taxpayers find that tax practitioners and tax advisers are indispensable guides when it comes to planning and filing their annual returns. In fact, given today's market, it would simply be unwise to pursue many personal business transactions without first consulting with a tax expert to determine the best way of arranging one's financial affairs.

It's understandable then, that as life and tax law become more complex, the tax professional has come to play a more important and ever expanding role in the American business community.

Tax Practice: An Overview

Income tax practice is an unusually satisfying, respectable, and profitable occupation. As a tax practitioner you perform a vital, professional function in the nation's economy and occupy a position of prestige and dignity. And whether you view this profession as a full-time career or whether you are happy supplementing your income by preparing tax returns during the tax season; you will, no doubt, find tax work to be stimulating, challenging, and enjoyable.

The tax practitioner is in a position to advise and assist people with many financial, and occasionally, even personal problems; saving them money, getting them refunds, devising tax saving strategies, and finding deductions they overlooked or didn't even know existed. Sharing in the most intimate facets of a client's financial life, the consultant becomes a trusted friend and adviser. Compensations like these are rare in any line of work, but they are your privilege to enjoy as a tax consultant.

Also, in the course of your practice, you will come into contact with individuals from all walks of life. As a result, you will acquire first hand knowledge of various trades, businesses, and professions; and gain insight into a wide variety of investment and other financial situations. The valued friendships, breadth of experience, and the detailed knowledge gained of business practices—all of these are of tremendous value to many practitioners in their business as well as in their personal lives.

Furthermore, it is very unlikely that tax work will ever become obsolete, and there is very little chance that the tax consultant will ever join the swelling ranks of technologically unemployed. 'Death and taxes' being the certainties that they are; an ever increasing need for tax consultants seems equally as certain.

And as far as financial rewards go—few professions or occupations provide a quicker or greater financial return for the relatively small amount of money, time, and effort required to enter the field.

Before going to the next chapter, however, let's briefly discuss a few commonly asked questions—some of which may have been troubling you too. First,the questions:

1. Isn't the computer making the services of tax practitioners superfluous? Why would anyone hire a tax preparer when you can get inexpensive tax preparation software and do your own return?

   Rest assured that in spite of all technological advances, preparing any but the most simple tax return by computer is still quite a formidable task for the untrained. The primary market for the low-priced tax preparation software is that group of taxpayers who have relatively simple situations and always prepared their own tax returns. And even many of these individuals, after a few disastrous attempts, give up and turn to professionals. They realize that the hassle of dealing with the complexities of the tax law and the often confusing computer programs just isn't worth it.

2. What good is my professional training? Why can't any one with a computer and tax software just set up shop and grab my clients, perhaps by offering lower fees?

   The answer to the previous question should allay this concern as well. Tax preparation software is an excellent tool—and like all professional tools—is worthless, or worse, if you don't know how to use it. At best, and if you are willing to spend the necessary time and effort, it may help you do your own return, but anyone who thinks that it can use it on a professional basis for different taxpayers and different situations is in for a rude shock.

3. And what about the large tax preparation chains and their enormous advertising budget? What chance do I have competing against them?

   There's no doubt that the large tax preparation chains—especially H&R Block—have taken over a large share of the market. However, their strength, i.e., quick, efficient, but impersonal, assembly-line tax return preparation by low-paid preparers with a high turnover rate, is also their greatest weakness. First, it pretty much restricts their clientele to the less lucrative taxpayer class, mainly lower-income wage-earners with fairly simple returns. Secondly, it makes it relatively easy to attract many of their clients by offering superior, personal service. Indeed, statistics show that the chains have not gained market share in recent years. Apparently, whatever gains they have made have come primarily from the natural population growth.

   Later in this book we will discuss in detail how you can capitalize on this inherent weakness and turn it to your advantage. We will describe proven methods and techniques that alert tax practitioners effectively use to cope with the competition. And we will show you how to build your business by weaning away clients from the chain tax services.

And finally...

4. I read that the coming FLAT TAX or other tax simplification measure will do away with income tax returns entirely, or—at most—require a 'post card' size tax return?

Have no fear! The 'Flat Tax" idea (and similar ones) has been around for several decades. Though it's politically appealing, it has gone nowhere and, according to most experts, is unlikely to be seriously considered in the foreseeable future. One reason is that to make it equitable and workable in our complex economy, Congress would have to enact so many new rules, exceptions, special provisions, etc. that the resulting tax system would be nearly as complicated as the current income tax.

As far as the siren call of 'tax simplification' is concerned, you know what happened every time in recent memory when Congress tried to simplify the law...it became only more confusing. In fact, the 1986 Tax Reform Act, which was termed at the time 'the most complex tax law ever" has become immeasurably more complicated as a result of all the 'simplification' amendments added since.

In summary, we feel that we can safely assure you that the independent tax professional is alive and well, and—with the help of Congress—doing better all the time. Based upon our intimate association with the field, we're convinced that tax practitioners will continue to prosper well into the future.

Tax Practice for Now...and for the Future

The tax practice profession has undergone some changes since this text first appeared in 1997, and for a number of reasons; among them being—frequently changing tax legislation, advances in technology and the ongoing information revolution, a more complex economic climate, and more complex business practices and procedures. Consequently, tax practice as a career opportunity has opened up and the demand for qualified tax practitioners is growing as well.

However, as the economic picture changes and as career prospects improve, new demands are being made on tax practitioners as well. And in order to be a successful, developing tax practitioner today, you must structure your career around something more than technical competence alone.

Of course, it goes without saying that technical skills and tax knowledge are absolutely indispensable to our discussion; without these essential skills there is nothing further to talk about. But as important as technical knowledge is, there are other skills you have to develop that are also very important...skills that center around your general business sense and overall effectiveness as a person. And this is primarily what this text prepares you for. Building, Marketing and Operating A Profitable Tax Practice may not make you a better qualified tax practitioner in the narrow sense of the word, but it will show you how to take the technical skills you have acquired—and continue to acquire—and convert them, employ them, channel them, in a way that can make you a more successful practitioner.

Building, Marketing and Operating A Profitable Tax Practice may not necessarily provide you with a better understanding of tax law and tax law application; for its overriding focus is to coach you in how to utilize your acquired skills to your best advantage. The text aims to help you effectively structure and manage your career. You will find chapters containing timely and practical advice on how to organize, manage, and communicate your skills; how to handle yourself and how to deal with others; how to promote your services, along with suggestions on how to make full use of your technical know-how—and all for the sake of successfully advancing your career.

Again—technical competence is paramount; but the fact that you are a qualified tax professional indicates that you already have the basic skills necessary for a productive career. To use a metaphor—your tax skills are like a well tuned engine; the task now is one of steering the engine properly in a desirable direction. And that calls for implementing the information gathered and presented in this text.

Basically, the picture that emerges for tax practitioners in the coming decade is of a practice built around more client contact; not only more contact, but more quality, multidimensional contact between client and practitioner. And this, in turn, calls for better communication and relationship skills: being able to inform, explain, and instill confidence in clients. Therefore, in addition to being technically proficient, you also have to be able to listen, manage, and advise as well. And in order to meet these challenges, we have revised, expanded, and updated our original text.

This text contains a wealth of information, advice, practical suggestions, and examples—but don't be overwhelmed by it all. It is not meant to be read and implemented in one night, or one month, or even a year. You can peruse the text, or read it from cover to cover to get an overview of the material; but it is designed primarily as a guide and reference work—to turn to as the need arises. There are chapters that may not—practically speaking—concern you right now. You may read them for their information value, and then study them more carefully when you are ready to actually make use of the information. In other words, the text is designed as a guide to actually make use of the information and as a guide to accompany you as your practice begins, changes, develops, and grows.

A Final Thought Before We Begin...

Some things, of course, change; but some matters of principle do not change. The following excerpt, taken (with minor changes) from the introduction to the first edition, still holds true today as it did then...

"...it is important to establish firmly in your mind that you are not a technical or robot trained to put the right number into the right space on the right form. Your duties, opportunities, and responsibilities as a tax consultant go far beyond the procedure of merely filling out a tax return.. If you really want to do justice to yourself and your clients you will apply your expertise, experience and ingenuity to assist them in paying the lowest legal tax possible. Also bear in mind that preparing tax returns means dealing with the lives of people, and not just with facts and figures. And finally, as you will be pleasantly surprised to discover, you will derive an enormous amount of satisfaction from delivering a professional level tax service. As the various forms and schedules take shape and fall into a pattern, you will begin to feel a sense of achievement—heightened by the fact that you have not only done your best to work out the lowest possible tax liability for your client, but have also helped him or her comply with the law. At this point you will begin to take pride in your work, in your respected and ongoing association with clients, and you will truly be glad you chose this honorable and satisfying profession."

One thing is certain: tax law is fascinating, intricate, and ever changing. And for as long as that remains the case, taxpayers—be they individuals or business firms—will be in need of professional assistance and advice to properly manage this most vital aspect of their lives.

Since this volume is intended to help the novice as well as the seasoned practitioner, we will now proceed to the basics of how and where to establish your practice. Those of you who are already in business may want to skip the next few chapters, but we feel a quick review of the material will yield some ideas as that every one of you will sooner or later find beneficial.

Chapter 2
TAX PRACTICE—WHAT IT TAKES TO SUCCEED

Success is discovering your best talents, skills, and abilities, and applying them where they will.

Wilferd A. Paterson

The reason most people do not succeed is that they will not do the things that successful people must do. The successful scientist must follow a formula. The successful cook follows a recipe... It is not important that you merely want to succeed, unless you want to badly enough that you are willing to do certain things.

Most people don't plan to fail... they fail to plan.

Author Unknown

According to the theory of aerodynamics, and as may be readily demonstrated through laboratory tests and wind tunnel experiments, the bumble bee is unable to fly. This is because the size, weight, and shape of its body in relation to its total wingspread make flying impossible. But the bumble bee—being ignorant of these profound scientific truths—goes ahead and flies anyway... and even manages to make a little honey every day.

Author Unknown

Today's Practitioner: What It Takes To Succeed

Introduction:

This chapter did not appear in the first edition, but we thought that before we begin to focus on outer furnishings and preparations—such as location, and marketing strategies, etc.—we should devote a few words to the subject of 'inner preparation'. In today's ever-changing and competitive marketplace, it's worthwhile developing a 'mindset' that promotes personal success rather than one that undermines it.

The chapter is far from exhaustive but it will highlight a few key attitudes and personal traits that are certainly worth mentioning.

The idea of running one's own business has long been part of the 'American dream'. In fact, in many ways America's 'entrepreneurial spirit' epitomizes that dream with its unique system of free enterprise; whereby anyone—with some capital, some skills, and some fortitude—can pursue any business endeavor that he or she desires. This is a noble freedom that millions of Americans continue to make use of. No wonder, 'going into your own business' is a phenomenon that is definitely on the rise.

Nevertheless, transforming this dream into a reality calls for the practical implementation of certain skills, insights, and attitudes which we will now discuss. A better understanding of some of these 'components of success' will be of help to you in your own development as a tax professional.

Nutritionists often say, "You are what you eat!" And if this holds true in a biological sense, then it probably holds true in a psychological sense as well... in that you 'essentially' are what you think and believe. In fact, this has proven to be the case: it is a well established fact that your attitudes—about yourself and about life in general—have a lot to do with your measure of success. By firmly believing that you will succeed, you improve your very chances at success. So to begin with; you have to believe in your own potential for success; in that you have what it takes to succeed. And even if you don't have 'all' that it takes just yet; you at least know that you can develop, acquire, or learn the necessary traits or skills that contribute to personal success. Everything that follows in this chapter comes on the heels of this fundamental belief in one's own potential.

Beyond this basic belief in oneself, those who go into business for themselves tend to exhibit the following attributes in some measure; and if not all of them, then at least some of them—depending upon what their particular task or work situation demands of them.

1. A drive for independence:

A desire for personal working autonomy coupled with an ability to work well alone—this seems to be an essential attribute of those who go into business for themselves. Of course, as a tax practitioner, opportunities abound for client and social contacts, but in a predominantly one-person-venture, you have to appreciate your independence while you effectively make good use of it.

We should quickly point out that independence here does not mean 'free to do whatever you wish'. It's independence from direct supervision above you; but you still have to make use of that independence to fulfill your various responsibilities. Therefore, an important corollary of this concept is that you ought to be a good 'self-starter'. Since you don't have your own boss watching over you; you have to be your own boss—to get up, get going, and get the job done!

2. A willingness to sacrifice:

Especially when you begin a new practice and are seeing if off the ground, you have to be willing to give it the time and attention it deserves. And this calls for hard and steady work. Of course, you don't have to abandon your other interests and associations; but you have to be willing to 'put in that extra hour' or 'go that extra mile' if it is necessary to do so. This is certainly the case when you are opening a new practice, and it will also hold true at various times of the year—such as at tax season and the weeks leading up to it.

3. A willingness to take chances;

Managing your own professional tax practice means that you must be willing and able to take occasional calculated risks. To be sure, there are a number of things you can always do to minimize the risks involved—such as planning, research, careful consideration of all essential factors—but as in any business, an element of risk remains, and you have to be able to deal with it.

Along with this trait goes the ability to make decisions. As a sole practitioner, you have to be able to gather the information you need and analyze your choices, in order to make sound and accurate decisions. Developed skills in this area will certainly help your practice to expand and grow.

4. An ability to persevere:

One of the most commonly cited reasons for failure in a one-person-venture is simply giving up too soon. You have to be persistent in your endeavors, and not let minor setbacks or negativity sway you from your goals. Naturally, all this is easier said than done, but the importance of perseverance cannot be overestimated. Don't let yourself be overcome by impatience or the fear of failure.

An extension of this principle is the idea that...not only do successful people persevere, but they also possess a healthy, productive outlook on failure as well. They tend to see failure in a more positive light: as an opportunity to stop and assess one's motives and abilities; as an opportunity to sharpen one's skills and one's resolve. A temporary setback can also be instructive, in terms of what to do and what not to do next. The ability to weather a failure can make you stronger, smarter, and better prepared for your next round in the business world—but all this holds true only if you are willing to examine the setback in this way.

Therefore, being successful means having the courage to meet failure without being defeated, and refusing to let a current loss interfere with your genuine long range goal.

5. A sense of enthusiasm:

Successful practitioners are basically optimistic and enthusiastic; not only about their work, but about life in general. They are, therefore, able to look upon their endeavors as exciting, adventurous, stimulating, and creatively challenging. This outlook, in turn, enables the mind to see the possibilities in every encounter; to visualize solutions to daily concerns; and to ultimately look upon problems and setbacks—not as obstacles, but as stepping stones—to further growth.

6. A willingness to learn:

Formal schooling is one way—but by no means the only way—to gather facts, accumulate knowledge, and gain insight into the ways of this world. But beyond any of the means at one's disposal, we are referring here to a deeper openness and willingness to learn and go on learning whatever is necessary to the proper running of your tax practice. Make use of community resources, libraries, lectures, seminars, updates. Be an avid reader—of books, newsletters, journals—so that you are in touch with current events, business trends, and financial affairs. We are dealing here with a healthy curiosity, with being a student of life, with keeping your eyes and ears open so that you can creatively learn all that you see and hear around you. As a result: the more you learn, the more confidence you acquire, and the better prepared you are to serve your clients—and thereby you enhance your chances at both financial and personal success.

7. An ability to plan and prepare:

Going into business for yourself—opening a professional tax practice—calls for planning, preparation, and a good deal of forethought. Planning and researching forces you to think, look ahead, and weigh alternatives.

You should have some concrete idea of what goal you are aiming at, and then do whatever you can in order to reach that goal. Your planning should take into account as many pertinent factors that you can think of (many of which are discussed in this text), that will affect your actions; and then with all your preparations before you like a road map, you are in a better position to advance into the direction of your own success.

You may compare this entire planning stage to the task of an archer aiming his arrow at a specific target. The clearer he is on the location of his target, the more skilled he is in the use of the bow and arrow, and the more carefully he sets about taking aim (considering all the conditions around him), the more likely he is to succeed at finally hitting the mark. The same holds true in your approach to your tax practice: with your skills in hand, with your goal in mind, and with proper aim and preparation—you increase your chances at success. And that is what this text is primarily about... increasing your chances at success as a tax professional.

Few people possess all of these valued traits, and fewer still possess them to any degree of perfection. Fortunately, however, as you develop your practice, you don't have to be the epitome of all these attributes. You only have to be willing to learn, cultivate them, and draw upon them as the need arises; and your determination to do what is right will help you and your practice to flourish.

The possibilities are great, the journey promises to be an exciting one, and what awaits you as you proceed is most exciting of all: an opportunity to live a flexible and expansive personal life; to play a larger, more influential role in your community; and to gain financial satisfaction and true personal fulfillment by contributing to your community in a meaningful way.

Chapter 3
ESTABLISHING YOUR OFFICE

A journey of ten thousand miles begins with a single step.

Confucius

Establishing Your Office

It is possible to operate a tax preparation office out of a briefcase, but if you are at all serious about getting into tax work on a more than incidental basis, you should have some sort of an office—even if it is in your own home—and acquire a minimum of office furnishings, supplies, and equipment.

This chapter will help you locate, outfit, and equip your office or work space.

Location:

Selecting a good location for your tax practice is one of the more important decisions you will have to make, because your location will strongly affect other aspects of your business: your operating hours, overhead, the amount and type of clients you attract, and much more. Therefore, it pays to do some careful research into the question of location. After all, it's a decision you do not want to make too often. A well considered location, in the long run, will actually save you time, energy, worry, and money.

To begin with, you may not even be in a position to select the 'best' possible location for your practice. You may—at first—want or need to work out of your home. Or perhaps, your limited finances will dictate the location of your office. And then again, you may be purchasing someone else's practice, and the transaction may include a predetermined location. So there are many factors that may automatically keep you from choosing an optimum location. But if you have any control over your location, and to the extent you do, here are some points to consider that will prove helpful.

1. Identify possible neighborhoods for your tax practice:

Begin by asking yourself: Who are my prospective clients and where are they located? You may identify this target market in various ways: by age, economic status, vocational characteristics; and your answer may include various types of people, but list them and determine approximately where they work and/or reside.

By the way, the study of populations and their make-up is a science of its own called 'demography', and a great deal of sophisticated statistical research data is available to you in this field. For example, a lot of big businesses study in detail the demographics of a specific region before opening an office or business there. And as a tax practitioner you can take advantage of all this available research when considering your own office location. If you want to pursue this avenue further, the following community resources are good places to begin:

- Your local library will have much of this information available as reference material, including detailed neighborhood maps from government agencies indicating population and business shifts and trends. For example, your library should carry Statistical Abstracts of the United States. This annual publication from the United States Census Bureau contains ample information on population density, growth, incomes, occupations, and similar data.
- The city, or town hall is a good resource, as well as real estate firms or banks.
- Your local Chamber of Commerce has data on different sections of the city and can be of help.
- Make a point of speaking to state, county, and town officials; people in business and business organizations. They, too, should know about business trends and the population potentials in a given area.

By the way, it's all too natural to think of starting your tax practice in your own home town, but you shouldn't assume that where you live now is necessarily your best choice for a tax practice as well. Your research may uncover some other pleasing and promising possibilities in other communities. Don't overlook these findings; check them out.

2. Other neighborhood considerations:

Once you have selected possible areas that represent a rather high concentration of your potential client population, you can then examine these neighborhoods from other worthwhile perspectives.

Also bear in mind that in determining location you are primarily trying to meet your clients' needs and not necessarily your own. For example, you may favor an office building perched on a hill with an impressive view overlooking the bay, but if it is of difficult access to most of your prospective clients, then it is of little practical value to you. In fact, try to think in terms of two major concerns when weighing any of the following aspects of office location: a) maximum convenience for your clients and b) maximum operating effectiveness for you...

- Traffic and accessibility—Is it a well trafficked area or an isolated one?
- Is there public transportation to where you are? Is the office easily reachable and easy to find, by car or on foot? On the other hand, is the area overly congested, noisy and dirty?
- Is adequate parking readily available? Is it free or is there a cost?
- Consider locating near compatible businesses and services; i.e. banks, real estate firms, insurance agencies, and other professional service-oriented occupations.
- Consider the competition: Is the area saturated with tax consultants, or is there room for one more? Or perhaps there are practitioners in the area, but the quality of the service is disappointing—leaving room for better service.
- Try to avoid blighted, run-down areas. Ask yourself: Is this neighborhood improving or deteriorating? How pleasant are the surroundings? Is the area clean or full of litter? Is the area well lit and safe at night?

3. The particular building and your office space:

After considering the population of the neighborhood, the prospects for business there, along with some of its physical attributes; it's time to examine the building and office space you intend to occupy. Here are some questions to ask yourself. Again, remember to view each office site as if you were a prospective client.

- What is the age and condition of the building?
- Is it clean and attractive, or shabby and hazardous?
- Is there adequate lighting outside the building and in? Are the grounds well kept?
- What kind of traffic patterns do you see around the building? What kind of people do you see in and around the building? • When you check out the office itself, consider...
- How many rooms or how much space do you have?
- What is the condition of the office's mechanical systems: electric, plumbing, heating, and air conditioning? Is there a restroom available? Is it clean and well kept?
- How are the floors, windows, walls, and ceilings? Check the acoustics. Is

remodeling or painting necessary? At whose expense?

- What type of lighting is provided?
- What are the terms of the lease, and what are the terms for renewal or cancellation of the lease?

In general, don't bother going into neighborhoods, buildings, or offices that are definitely unsuitable. And unless you need an office immediately don't settle on the first site that seems adequate. Shop around and try to find an office that has as many positive features and advantages as possible.

Opportunity for expansion:

It may seem premature to you now, but if you use common sense and follow the guidelines and suggestions presented in this volume, you may soon find yourself in need of more space. Moving an established business is always costly, traumatic, time consuming, and may even result in the loss of some clients. Therefore, see if you can make some provisions in advance, in order to avoid—or at least postpone—the necessity of relocating.

For example: Try to initially obtain more space than you need. This may be expensive; but if you are renting an office, your landlord might be willing to insert a clause in your lease giving you the option to take on additional, adjacent space when it becomes available. Or, you may be able to lease more space then you need, and sublet it until you actually need to make use of it.

If you build an addition to your home to house your office, you can probably incorporate provisions for future expansion into the current plan, at little extra cost.

As a general rule, try to envision your business needs three to five years from now, and see if you can obtain a space that is both adequate now and that can also accommodate your future needs.

Checklist:

What to Look for and Ask About When Office Hunting

- Access to building—mass transit, other; parking facilities
- Access within building—elevator or stairs
- Condition of building—inside and out
- Office space—number of rooms or size of available work area
- Number and location of electrical outlets/type of wiring

- Terms of lease and monthly rent—length of lease and conditions of lease
- Are any repairs or alterations called for—paint, doors and locks, security, floors and carpets, windows and walls, ceilings, drapes and curtains, etc.
- Who pays for maintenance, utilities, repairs, and other services (cleaning, trash, etc.)
- Are office signs allowed—inside and out
- Traffic conditions
- Noise levels—inside and out
- Restrooms—location and conditions of
- Exits and/or fire escapes
- Lighting situation—outside of building and inside as well as inside the office: Is it sufficient? What type is it?
- Heating and cooling systems—type of system; is it working?
- Office Arrangement
- Other features: ______________________________

For a tax professional, creating an office environment that is effective, efficient, yet congenial, is an important consideration. Your office should be set up in a way that enables you to get your work done; at the same time the arrangement should allow clients to feel physically comfortable and psychologically at ease.

To operate effectively most practitioners need a work area that can accommodate...a desk and chair from which you can do paperwork and make phone calls; space for some measure of electronic equipment, be it a typewriter, computer, or other form of technology; space for storage—a filing cabinet, bookshelves and reference materials; and some space for storing supplies and other equipment (this may even be in a different room).

If you have more space to work with, you can designate it in accordance with your needs: a conversation or meeting area; a waiting area for clients; or a work space for such activities as assembling, producing, or mailing materials.

Of course, all of these considerations depend upon the amount and configuration of your available space. If you only have one room to work with and you need a waiting area, then room dividers, partitions, or screens are helpful; even bookcases and file cabinets can help section off one area from another. And while room dividers can provide some degree of privacy, curtains can help close off storage areas where you keep supplies and accessories that don't fit into cabinets.

One possible solution to layout problems in tight quarters is movable furnishings. Many furniture units—chairs, files, desks—come on casters or locking wheels. This allows you to easily transform your room arrangement contingent to your needs. You can keep computer, printer, copier, or typewriter on a movable stand and roll it away when not in use; or if you have a lot of books you can use tall bookcases and make better use of vertical space. Furthermore, manufacturers today are designing office furniture and equipment that will enable you to make optimal use of your limited space. For example, there are compact office file cabinets that fold out to become a full desk, and that can roll into any corner of a room.

In any event, when it comes to designing and arranging whatever space you have, you should try to strike a balance between practical and aesthetic considerations; bearing in mind—personal preferences, client needs, as well as your professional goals.

Office Decor

As you plan out the interior of your office, think of making it as pleasant a setting as possible—for yourself and for your clients. Pleasant for yourself means a setting wherein you can effectively and efficiently spend many hours of time; and for one thing, this means cutting down on disturbances: excessive noise, movement, and poor lighting.

Noise Control

There are a number of things you can do to make your work area more effective—from a 'sound' perspective:

- Carpets, curtains, and drapes are not only attractive, but they can help reduce extraneous noise. Furthermore, they are excellent insulators; keeping out both Summer heat and Winter cold.
- High ceilings from which sounds reverberate can be lowered with acoustical tiles or other sound absorbent material.
- Weather stripping on doors and windows, padded furniture, cushions and rubber pads under office equipment, and room dividers can also help reduce noise.

Some people find that not only is too much noise an interference, but that too little is also disturbing. If you fit into this category then you might consider installing low playing music, a fish tank, or songbirds to alleviate the situation. Such 'natural' sounds may help you to work better, make your office more pleasing, and may also help screen out noises that can't be masked in any other way.

Minimizing Movement

The best way to handle disruptive movements—people coming and going—

is to be in a separate room. If this is not possible then decorative room dividers may be helpful. And if this is also not a practical consideration, then try arranging yourself and your furnishings in such a way that you don't have to look up every time someone passes by.

Lighting

Lighting has a lot to do with comfort and productivity: it reduces fatigue, promotes accuracy, and can provide a cheerful working environment; so you should consider this matter carefully.

The best light for your office, of course, and the cheapest, is natural sunlight—through windows and even a skylight, if possible. It's impractical, however, to rely entirely on daylight, so you have to supplement with artificial lighting.

- For more general illumination you should have some form of overhead fixture—either fluorescent or incandescent lighting. Each has its distinctive advantages, and you will have to assess which works best for you.
- For more task oriented lighting—to get specific jobs done—you should have some sort of adjustable—desk, swing-arm, or floor lamp—that you can use for more focused lighting needs.
- With whatever lighting you choose, be on the lookout for eyestrain glare, or other light-related conditions, and try to make adjustment accordingly; the light may be too weak or too strong, you may need frosted bulbs instead of clear, or better shading, or you may not react well to fluorescent lights. (Bear in mind that people working on computer screens need subdued, glare-free light).

Walls and Accessories

Office walls can be functional—with calendars, bulletin boards, notices, and news clippings; or decorative—with paintings, pictures and tapestries; or both—depending upon what works best for you. Of course, with whatever you put on your walls, try to maintain a professional posture.

If you have to refinish walls—paint, wallpaper, or paneling are among the more common ways of changing the appearance and character of your wall space. Just bear in mind that the colors you choose can have an impact—both on mood and work efficiency. Basic, traditional, earth-tone colors are more conducive to a professional setting than loud, vibrant colors and patterns. Use bright colors mainly to accentuate your surroundings.

The colors and color combinations you employ should be pleasing, and yet enhance your ability to function.

Furniture and accessories should also be tastefully selected—with concerns of comfort and professionalism in mind.

- In addition to whatever essential furniture you need, the addition of some natural elements is always appropriate: plants, flowers, dried flowers—even a fish tank if there's room.

- Paintings, prints, or photographs (but not too many) can do a lot to enhance the appearance of a room. Other wall decorations may include small rugs, tapestries, or even a mirror.

- Displaying evidence of your professional accomplishments, competence, and expertise; certificates of achievement; other awards and credentials; or expressions of interest and personal concern—all of these are certainly appropriate as wall, desk, or table ornaments. This could include professional licenses, diplomas, trophies, a family photograph, or social involvements—i.e. a photograph of a little league team you help to coach. Not only do such items help personalize your office, but they instill clients with a sense of confidence—in you, your abilities and values.

- If you have a separate waiting area, be aware that this will be the first visual impression visitors will have of your office procedure. Have some comfortable furniture available, and a magazine rack or table with a variety of interesting reading material. Make sure the lighting is adequate, and have some interesting things on the walls—even a decorative clock. Waiting won't be such an irritating experience if there is at least something interesting to read or look at.

- Remember, too, that even if you have no more than a small work station to operate from...make the most of it: personalize it, decorate it, make it as welcome and as comfortable an area as you can; so you can effectively spend as much time there as is necessary. Too often, people make the mistake of only seeing the limitations of their work area and fail to modify it to meet their working needs.

Selecting Office Furniture

When you look for office furniture try not to get carried away by gimmickry and sleek design. Examine a piece of furniture and ask yourself these questions: Is it practical? Useful? Durable? Easily maintained? Is it worth the expense? Will it help me perform better at work?

Desks and chairs:

It's no exaggeration to say that you may be spending more than half of your working day in your chair—at your desk—so these are items that must serve you well.

Look for a sturdy chair that gives your body as much natural support as possible; your chair should be comfortable and encourage you to sit with good posture. A high quality, adjustable posture chair can go a long way to reduce fatigue and increase productivity. With adjustments in seat height, and back height and back tension, you can perform a variety of tasks without inviting unnecessary aches and pains. Padded seat, back, and arms; being able to swivel around; and strong casters that enable you to roll around from one area to another—these are further considerations in the purchase of a good desk chair.

Since you spend so much time in your chair, it pays to put some extra money there as well. If you skimp on your chair and sacrifice your health, the money may just go to doctor bills instead.

By the way, since the human body was not meant to stay put in one spot for so long, it's a good idea to periodically get up, walk around, and get some fresh air and sunshine.

Choose your desk with function and practicality in mind. Desk tops tend to be loaded down occasionally—not just with papers but with heavier equipment—so your desk should be sturdy and able to carry the load.

To be normally functional your desk should have a file and supply drawer, and the drawers should lock.

Depending upon how much space you have, try to get a desk with a reasonable surface area so you can sort things out and have enough room to maneuver.

Storage Systems:

A good filing system is essential to good office organization. You will almost immediately have to 'put things away' and you will have to know where these things are so you can easily retrieve them. Therefore, a good file cabinet is an early purchase for your office. It doesn't matter if it is a two or four drawer file; just be sure it's sturdy, with drawers that slide easily, because it undergoes a lot of steady use. Buy an ample number of hanging files and manila folders; and since you carry confidential material, locks are essential.

For additional storage space, you may want to consider buying a credenza to place behind or alongside your desk. This is a useful piece of furniture equipped with file drawers, box drawers and shelves; and with casters, you can easily roll it from place to place.

If you have subsidiary items that you need only rare or occasional access to, then cardboard or plastic archives boxes may suffice. And what does not fit into a filing drawer or cabinet may go on a shelf. Making good use of vertical shelf space is especially effective in a small office setting.

Depending upon your office needs and what space you have available, here are some other furnishings to consider:

- Additional tables—Tables (even folding tables) can add to your surface working area, can support equipment, and even double as a conference table if need be.

End tables or coffee tables can enhance your waiting area;

- Extra padded chairs (or even sturdy folding chairs) in case you have a conference or meeting;
- Wall clock and mirror;
- If you don't have a closet you will also need some sort of coat and hat rack where these and other personal client accessories can be stored.

Office Supplies:

Whereas office furnishings and other equipment are, for the most part, a one time investment, office supplies always need replenishing and are therefore a more constant drain on your operating budget; plus—office supplies tend to be expensive. Therefore, it pays to shop for these items with thrift in mind:

- Be sure to compare prices. There can be a significant difference in prices between one office supply store and another. And very often a discount outlet will carry just what you need—at substantial savings.
- Office supply items are generally less expensive when purchased in quantity; so if there is something you know you will be making constant use of, try to buy as much of it at one time as you can.
- Keep an eye out for quality as well. Try to buy items that are durable and that will serve you well. A cheap stapler that will soon need a replacement is not a source of savings. On the other hand, if you need pads for scribbling lists or notes; there's no need to buy high gloss, heavy duty paper. So the quality of an item can reflect its use or function.

Checklist of Office Supplies:

You may or may not need all of the items listed below but it is a basic list of common office supplies—items that most office personnel tend to use at one time or another.

- Accordion files
- Bulletin board and push pins
- Business cards
- Calendars: desk, wall, datebook, or appointment calendar
- Correction fluid
- Dictionary and Directories
- Envelopes: small, business size and large manila envelopes
- File folders, labels, and tabs
- Glue or rubber cement
- Hole puncher
- In/Out trays
- Index cards

- Paper: pads, message pads, "Post-it"notes, notebooks, and computer or typing paper
- Paper clips
- Pens, pencils, pencil sharpener, erasers, and markers
- Postage stamps (unless you use a postage meter)
- Rolodex type—name, address, and telephone number file
- Rubber bands
- Rubber stamps
- Rulers
- Scissors
- Stapler, staples, and staple remover
- Stationery: letterhead and billhead
- Tape and tape dispenser
- Typewriter ribbons
- Wastebasket and trashbags

Tax Forms:

Absolutely essential to your list of office supplies is to be well stocked with a broad array of tax forms.

If you prepare your returns by computer (see discussion later in this Chapter), your software package will generate most of the tax forms and schedules you need. If you prepare returns manually, you'll need tax forms. Since IRS no longer supplies forms in quantities, you can order a few copies of each of the major forms from IRS and reproduce them on your copier, as you need them, or purchase them in bulk from one of the accounting supply houses. If you rely on the IRS, make sure to get your order in early and check carefully to make sure that you received everything (errors and omissions are not uncommon).

However, as your practice grows, you'll often need some of the less common forms and schedules—forms your software does not include or that you did not order from IRS (or that you haven't received) because there's no way to anticipate all your requirements. While you can order the desired forms from IRS as you go along (and hope you'll get them in time) not having the right forms, or instructions, when you need them can tie your office up in knots. There are several ways to handle this problem:

1. If you're properly equipped you can download them from the IRS Internet website. But it is a rather tedious process, and during the busy tax season you may have trouble getting to the site.

2. If you have a computer with a CD-ROM drive, your best bet is to purchase a CD-ROM tax forms service (not to be confused with a tax preparation program) that'll give you just about every tax form—with instructions—in existence. The

better services (including the **National Tax CD-ROM** service) are frequently updated during the tax season and give you the choice of printing out blank forms, or filling them out on screen.

3. If you are not computerized, or—like many—prefer to work with paper copies, you can purchase our **Master Federal Tax Forms File** and/or the Allstate Personal State Tax Forms File, each containing several hundred tax forms and instructions.

Basically, the same applies to state tax forms, except that these forms—especially from out of state—are unfortunately harder to get. Many states are notoriously slow in shipping forms, and a growing number charge for them. And even if you are Internet-equipped, not all states have web sites—and many of those that have sites don't make their forms available until late in the season.

So if you do any quantity of state tax work you should have either a CD or a paper-based tax forms service.

For information on **National Tax's CD** (which includes all federal and state forms and instructions on one disk) or paper forms services, please call 1-800-914-8138.

Office Technology

With office technology advancing as rapidly as it is, there is currently a lot to choose from in the way of mechanical equipment and devices that stand ready to make your office work more streamlined and efficient. However, not everything that is out there in the marketplace is necessarily for you. You—for your part—must be ready to evaluate your position, financially and otherwise, to determine what your technological needs consist of, and then see to what extent you can comply with those needs. Once you become aware of your position, you will be able to take the steps that are economically feasible for you, and that will enable you to operate optimally in your tax practice.

What follows is a brief description of some of the more basic technology to be found in many business practices today.

The Telephone:

For most businesses the telephone is a primary source for transmitting and receiving information. And in addition to standard telephone features, there are a variety of other functions currently available that may be of value to you—depending upon your individual needs and circumstances. In fact, the technology changes so rapidly that devices that seemed to be luxurious just a few years ago may now qualify as basic necessities, both in terms of professional image and personal efficiency.

If all you currently do is place and receive occasional, routine calls, then a standard phone may be enough for you. But if your telephone contact is extensive, than you may want to consider some of these additional features:

- Call forwarding allows you to automatically relay incoming calls to any other preselected telephone number.
- Call waiting signals you when another caller is trying to reach you while you are on the phone.
- Three-way calling permits you to talk to two people simultaneously while conference calls allow you to talk to more people at the same time.
- Speed dialing enables you to program frequently called numbers into the telephone.
- A 'mute' button keeps people from listening in on an extension.
- There are also 'speakerphones', and phones that keep track of the date, and that can measure the length of each call (that may be useful when charging clients for long distance calls).
- If you are on the phone a lot, you may consider using a lightweight headset; that way your hands are free to do almost any kind of work.

Telephone answering machine:

Just short of having your own secretary—and a mainstay in almost any growing business concern—is the telephone answering machine. It is your best option for providing telephone service when you are not available.

There is a wide range of answering machines on the market today, with the price dependent upon the quality of the merchandise and sophistication of its functions. And here again, the features to shop for really depend upon your needs. For example, there are machines that only give out messages without receiving the caller's information, and there are machines that record incoming calls—allowing the speaker to talk as long as necessary. There is also 'remote paging' where you can retrieve your messages from any touch-tone phone.

When you finally decide upon a phone system or individual features to purchase, do look for durability and reliability. In the long run it will save you money. Avoid telephones and telephone systems that are obviously 'cheap'; they sound that way and they detract from your professional image.

Telephone Courtesy

Very often, the telephone is the first—and only—way a potential client has of judging you and your practice, so be sure to employ telephone courtesy at all times.

Also bear in mind that the telephone calls for communication skills that differ somewhat from those used in face-to-face contact. With a telephone you must convey your attitude—your sincerity, and sense of confidence—through your voice. Here, then, are a few points to consider in order that you can make wise use of this vital instrument.

1. During business hours, the telephone should be answered—by yourself, a secretary, or an answering machine—every time it rings. While on the phone, be pleasant but businesslike; relaxed and sincere.
2. A business conversation should be free of all background noise—even music.
3. When preparing for a call try to have all pertinent information at your fingertips: client files, calendar, appointment book, etc.
4. Consider carefully how you answer the phone. Do you just say "hello", and then wait for the caller to respond. This may suffice in your home, but for business purposes it is not enough. You must identify yourself. Begin with an appropriate greeting—good morning or good afternoon—and then state your name and/or company's name.
5. Don't leave a caller 'on hold' for more than half a minute, if you can help it. If necessary, arrange for the person to call back, but don't leave a caller hanging indefinitely and without a reason. It's irritating, and it may even cost you a client.
6. When giving a caller pertinent facts and figures, speak slowly and distinctly—and speak into the telephone. Many are in the habit of cradling the phone in such a way that it juts into the speaker's neck or chin. This muffles the sound and can easily frustrate the caller.
7. Be aware of the importance of language. Over the phone a client only has your words to go by; so be careful of both what you say and how you say it.
8. Listen carefully and effectively. Listen to what is being said and how it's being said, and don't interrupt the caller unnecessarily. You can even reinforce the client by acknowledging your understanding of his remarks in some way.
9. Don't hang up on your client. Wait until the other party hangs up and then proceed to do the same.
10. In terms of both time and money, the telephone is your most effective communication tool: exceptional for scheduling appointments, sharing information, answering questions, and settling matters. Learn to use it well and use it wisely.

Typewriters:

In the 1980's it was thought that word processors and computers would render the typewriter obsolete. But despite such predictions, the typewriter looks like it will stay on as a standard piece of office equipment. For one thing, not every office or business needs a computer; and not every task calls for a computer. There are still plenty of jobs—memos, forms, proposals, envelopes, letters, and other correspondence—for which a typewriter is sufficient.

In fact, if you prefer the 'typed look' for your tax forms and don't prepare them by computer, a typewriter is much handier than a computer printer or word processor.

The current electronic typewriter is actually a cross between a typewriter and a computer, and depending upon its level of sophistication, can incorporate a wide variety of automated features. In fact, there are new, innovative machines appearing all the time within a broad price range. And with automatic features, display screens, and memory and storage options; it is sometimes difficult to distinguish these typewriters from certain personal computers.

As with all other office technology, new models appear regularly and prices of older models drop drastically. You will have to assess what your practice needs, and respond accordingly. With a little patience, however, you will see that equipment with the features you want will probable fall into your price range...sooner than later.

Copiers:

A good, reliable copier is probably one of the first pieces of equipment you'll want to invest in.

With copier capabilities on the rise, with more models available, and more sizes to choose from, the cost of reliable copy machines has come down considerably; so you do have an increasing number of quality choices in your favor.

For example, most manufacturers—recognizing the increasing demand for lower price, home sized models—are producing smaller, more compact copiers. These machines are nearly maintenance free; you can forget about powders, toners, drum cleaning and drum replacement. All you do is snap in a new cartridge after every few thousand copies. And not that you need it, but you may even find that zoom, reduction, and enlargement capabilities come as standard features on such equipment.

Printers:

If you have a computer, you have to buy a printer to go with it; and here too, there are vital choices to be made—from dot matrix to ink jet and laser printers, and varieties in between. Lasers, generally, are tops in print quality and speed. Because they have few moving parts, they rarely need repairs. Lower-end models can now be had at very affordable prices

Aside from financial consideration, the decision of which printer to use should be based upon what you want to print and how you want it to look. In any case, it is important to exercise care in selecting a printer that is sufficiently compatible with your computer and software.

Here again, the technology is changing rapidly and the prices keep coming down; there are a variety of compact printers for homebased offices, and even typewriters that double as printers—and that may be just what you need.

Fax Machines:

Except for Email facsimile machines are the fastest, most convenient, inexpensive, and by far—the easiest way to send documents, pages of text, and even graphic material—from one place to another. But here too, acquiring a fax machine depends upon the extent of your practice and your needs. However, as more fax models appear and as low cost units become more available it pays to shop for more than just the most basic model. Since you're dealing with documents that you may need years later, a plain-paper fax machine, in particular, is worth the extra money. Regular, thermal-fax paper copies fade fairly soon.

The most popular models combine the convenience of fax and phone functions—all in one unit. And if you have very limited desk or work space, there are models that combine, fax, phone features—all into one compact high-tech unit.

Dictating or Recording Equipment:

If you have a fair volume of business that involves a steady flow of correspondence, then a dictating machine would enable you to delegate this task to a part-or full-time secretary/assistant, who could then formally transcribe the information. For occasional work, an ordinary cassette tape recorder will do, but once you get into heavier volume, you should consider a regular office dictating machine.

Such equipment can be expensive but it's an investment that can long endure, and save you considerable time as well. And once you have it, you will find many other uses for it... memos to clients, bills, and recording information received over the phone.

The most important feature to look for in dictating equipment is sound fidelity; your voice should reproduce clearly so that a secretary can understand what is being said. This is especially important when your work involves the careful and exact transmission of numbers, specifications, and other vital financial data.

A good desk-top playback machine has a variety of features to make transcribing easier: simple push button operation, plugs for attaching foot pedal controls, and earphones.

Postage Scale and Postage Meter:

Trips to the post office are unavoidable when operating a tax practice; but there are ways to help you get your mail out more efficiently and promptly, and that can also save you money on unnecessary postage.

Since you'll be sending a lot of different-weight mail, a postage scale will take the guesswork out of calculating the correct postage. It'll make sure that you don't overpay, and—even worse—underpay and run the risk of having critical mail returned for insufficient postage, or risk the embarrassment of forcing a client to make up the shortage.

A postage meter is a handy device. Besides giving your mail a more business-like appearance, it'll also save you hours of standing in line at the Post Office.

Calculators

In addition to a heavy-duty printing calculator that you'll need, we recommend an inexpensive nonprinting desktop unit with large keys for quick figure-work. And make sure to carry a small hand-held calculator in your briefcase, purse or pocket at all times.

Computer Technology and the Tax Practitioner

Computer technology continues to have an immeasurable impact on all forms and facets of the business world, and this includes the field of tax preparation.

It would be impossible to cover, in this text, the subject of computers as it relates to tax practice; primarily because the technology is so extensive and it changes so rapidly, that whatever we say would be incomplete, and perhaps obsolete, even before we got to print. So to give detailed advice would be impractical. Nevertheless, by way of an introduction, certain guidelines are in order, and we will share those with you now. The objective here is to merely introduce the material to you, convey a sense of the possibilities that exist, so that you can investigate the subject further on your own and consider what might be the next desirable step to follow—for yourself and for your practice.

A computer can carry out a wide variety of office functions, and seems to operate best at those tasks a person finds most tedious and time consuming. A computer can perform these functions more rapidly and with greater accuracy—from maintaining accounts and files, budgeting, typing personal and business letters, addressing envelopes, and of course, tax research, plus a great deal more. And the time you save from these jobs can be spent doing more creative, productive work to advance and enhance your practice.

So the very first thing you have to do is sincerely determine just what your computer needs are. Precisely what do you want a computer to do for you? What general office tasks, and what specific—tax related—office tasks? And then find the software that does just that. Then, of course, you have to purchase the hardware that can accommodate that program or programs. Simply stated: Once you have determined that you are in a position that calls for the purchase of a computer system, then you must select the software and hardware required to perform the tasks you have in mind.

All of this, however, is easier said then done. Since a computer system is a major decision in relation to your tax practice, and since it calls for a considerable investment of time and money; it is to your advantage to think clearly into the matter and examine your possibilities well. Before you reach any decision, be sure you...

- consult with friends, relatives, or fellow tax practitioners who are familiar with the technology;
- do some reading on your own—books, magazines, journals, newsletters;
- take a course in computers or attend a seminar—before you purchase anything;
- consider hiring a well informed consultant for a few hours of professional advice.

Only after careful study and research should you finally select and purchase your equipment. Remember, too, that once you make your purchase—in all likelihood—you still won't be able to sit down and get to work. To effectively use your computer system you will have to spend some time in training to attain a certain level of proficiency. For training opportunities you can check the business section of your local newspaper for specialized courses and seminar; or contact the following resources: local colleges and universities, technical schools, business schools, courses offered by computer dealers, computer consultants, computer manuals—and don't forget to speak with relatives and friends who are in the field.

One caveat: For general tax and accounting work you don't need to invest megabucks in the very latest, state-of-the-art, lightning-speed system with all the attendant bells and whistles (which will anyway be outdated shortly). The best time to buy, usually, is shortly after the major manufacturers announce a significant upgrade and the dealers offer deep discounts to clear out their current stock.

Tax Preparation Software:

Income tax returns can be prepared manually or by computer.

Preparing taxes manually does not call for investment of time or money into the world of computer technology, but preparing returns by hand is slow and there is a greater likelihood of error given the complexity of today's returns.

The overriding trend, therefore, is definitely toward computer tax preparation. In the last analysis, this will save you precious time, and increase your efficiency, accuracy, and productivity. It also allows you to spend more time developing your practice—both qualitatively and quantitatively.

There are currently more than a dozen professional tax return software packages available, and though they have most fundamental features in common, they differ in features, price, ease-of-use and other distinctive ways as well. What it all boils down to is...that each practitioner must take the time to evaluate and compare various software programs to determine which would best suit the needs of his or her practice.

Points to Consider Before you Purchase any New Technology for your Tax Practice

Matching office technology to suit the needs of your practice is a complex undertaking.

First and foremost, consider the acquisition of new technology for your practice only as you genuinely need it. Don't make the mistake of buying more equipment, accessories, and options than you actually need. And to help you determine whether you need it or not, you might ask yourself the following questions: Will this piece of technology help reduce my expense? Will it help increase my income? Will it save me time and help me increase output, accuracy, and efficiency?

Furthermore, to evaluate the true cost of any piece of technology—high tech or low—you will want to know more about supplies, accessories, and their availability; service and maintenance costs; the costs of special features, special furniture that might be needed, and additional expenses. For example, with a computer... how will you be trained? Does the price include training? Is there a warranty, and what does it cover in terms of time, parts, and labor? All of these factors feed into the true cost of any technology you purchase.

Some experts in the field also suggest that individuals going into business for themselves should spend no more than one to five percent of their gross income toward the purchase of any new technology in any given year.

One other consideration: there is a lot of sophisticated technology available today and there is a lot of market competition—which is good for the buyer; but don't purchase without first comparing prices and features. And try sticking to a manufacturer with a good reputation for quality merchandise, reliability, and that offers the best service contract and warranties.

An Office In Your Home

Home based business ventures are definitely on the rise, and for a variety of reasons. For one thing, so many changes have occurred in contemporary society that have made working at home a great possibility: the computer revolution; the rising costs of office space; the increasing amount of time and money spent going to and from work; and a shifting emphasis in our economy to one of service and information exchange.

Of course, working at home is still not for everyone. Whereas an office is strictly a work area, a home is primarily a shelter for you and your family. The question is: If your practice is at home, how will you handle the many activities that take place on the premises that are not business related? If you're the type that can develop the discipline and the strategies necessary to handle the distractions, then working at home may be good for you. But if you cannot clearly separate your work from your home life, then you may be better off in a straightforward office setting—away from home.

Some advantages to working at home...

1. Financial—You can save money on commuting, fuel, parking, office expenses; and there are tax deductions too.

2. Time—You can save time preparing, waiting, commuting; and the hours saved can be spent productively in your practice.

3. Flexibility—In terms of life style and commitments, you can be more flexible in how you manage your time; working as early or late as you wish, and taking breaks as you need them.

...and some disadvantages—

1. Interruptions and distractions—Both home and family needs, the expected and the unexpected, may keep you from giving your practice the time and attention it deserves.

2. Loneliness—Many people need a work environment around them—offices, workers, the hustle and bustle of a business atmosphere—to bring out the best in them, and they just don't function well in isolation.

3. There are others who may find it counterproductive to be at home 'all the time'—both for purposes of living and working. It may also put a strain on family relationships, which can be a serious problem—especially if both spouses work at home.

If you do opt to operate out of your home, here are some tips to help you along...

1. If possible, try to keep your office or work space physically separated—even

closed off—from your living area.

2. Try not to use your office space for anything other than your tax practice.

3. Keep regular office hours, and get into the routine of treating your day in a professional manner.

4. Take breaks as you need them, and be in touch with professional colleagues.

5. Politely inform or re-educate family, friends, and neighbors, regarding your new routine, practices, and policies.

6. If possible have a separate business phone for use during business hours. Use an answering machine during non-working hours.

7. Although you can certainly personalize the decor in your home office; you should still try to maintain an image of professionalism, so that visiting clients see that you take your business seriously.

Full Time or Part Time?

One other matter worth mentioning that might affect your choice of office arrangement is the degree to which you are committed to your practice.

Generally, tax practitioners fall into 3 categories:

1. There are independent tax preparers who work part time during the tax season, though they may do some individual tax work throughout the year.

2. Others work full time during the tax season, and continue on a part time basis the rest of the year.

3. The third group consists of those who establish themselves on a full time basis throughout the year. Since there is rarely enough individual tax work available during the off season to be fully occupied, these practitioners devote the rest of the year to business tax returns, financial and tax planning, and related work... bookkeeping, social security practice, and similar activities. We will discuss this in detail in a later chapter.

The question is: To which group do you belong? And to a great extent the answer to this will depend upon your current situation plus your future plans and aspirations.

If you are now fully and gainfully occupied in a job or business, are pleased with your work and earnings, and future prospects are good; then you probably look to tax work primarily as a second income—a lucrative source of seasonal, part time earnings. The same reasoning would presumably apply if you are an active homemaker with some time to spare. Or perhaps you are semi-retired and are looking to creatively occupy some part of your day.

In these and in comparable situations, you would probably fit into the first two categories mentioned above. It is also quite possible that you are in a period of transition, trying to shift from the first two categories into the third. If, in fact, any of these descriptions relate to you, then you might want to look more seriously into one of the following office arrangements.

1. A home-based office (as described above) to begin your practice—provided that your home is easily accessible to clients.

2. Going to your client's home or place of business: Very often clients are willing to pay a higher fee for the convenience of your visiting them, and it is often advisable for the beginner—breaking into the field and acquiring a clientele—to employ this method. Many large and successful tax practices have started this way.

   In fact, visiting clients has taken on a new slant in this age of computers. If you are proficient with computers, then lap-top or notebook computers that weigh in at just under 6 pounds are worth your consideration. They come with handles, are portable, can fit into a briefcase, and can probably run any of the software programs you currently use. This new and exciting category of office technology can turn client visits into a busy and lucrative practice.

3. Office rentals: You can rent a vacant store or office with a street entrance, in a good business district—preferably in an area with a lot of pedestrian traffic. A store or office in a shopping center is also an excellent choice. Such locations can attract a lot of walk-in clientele. However, this is not always a practical choice for the beginner because of the high rent such prime locations usually demand. Furthermore, if you expect to be open during the tax season only, it may be difficult finding such a location, and you may have to look for new office space every year.

   As your practice grows however, it may be worth your while to pay rent on such space for the entire year, just to keep the office on a permanent basis and to remain clearly visible to the public. You may even be able to sublet the office, during all or part of the off-season months, in order to cut your expenses.

   Another option is to rent a small office in an office building. Rent is bound to be considerably lower, but you may get very little walk-in business. Here again, the same off-season drawbacks apply. But if you do rent in an office building, make sure that you are able to display a prominent sign, at the building's entrance or in the lobby.

4. Renting desk space: You might also consider securing desk space in a store or office; preferably in an insurance, real estate, or similar office setting. In most cases such arrangements work out to the mutual satisfaction of all concerned. Just make sure the office is located in a good business area, is visible to the public, easily accessible from the street, and that you can make use of the space

evenings and on weekends. Most part time practices operate evenings and weekends because clients generally prefer to meet after business hours when they are more at ease. Also consider your need for privacy in such an arrangement; so that you can engage in confidential discussions—if you do meet with clients during business hours.

Leasing your own space in a larger office setting- but using central facilities and services—has recently developed into a popular concept in office design. Tenants pay a proportionate share for a receptionist, secretary, and other services used. Very often such centers provide access to copiers, printers, fax machines and other forms of office technology. Again, you pay proportionately for what you use. Check into the overall cost of such services; but such a setting may be an ideal solution for a practice just getting off the ground.

If you happen to be 'in transition'—moving your way up from a part time into a full time practice—we suggest that you proceed gradually. Rarely, is the average beginner fully occupied in the first two or three years. Tax practice, like almost any other profession, is a growing, evolving business. It takes time and perseverance to develop a lucrative practice; though it is time well spent when you consider the invaluable experience you gain while building up your practice.

The information above is merely intended to provide you with a general outline of what you can expect as you begin and develop your practice. However, you may find that the pattern varies in your own personal circumstance. But above all—do not be discouraged if business is slow at first. You are sure to become busier as the tax season progresses, as your reputation grows, and you become better known—both in your neighborhood and in your community at large.

Chapter 4
SUCCESSFULLY MARKETING YOUR PRACTICE...

He who has a thing to sell, and goes and whispers in a well, is not so apt to get the dollars, as he who climbs a tree and hollers!

Anonymous

By Way of Background

Not long ago—in fact, as recently as 1977—a chapter on marketing in a text for tax practitioners might have seemed totally out of place. It was considered inappropriate for those involved in 'professional services' to go out and actively solicit clients. You could certainly develop 'a practice from within'—through the traditional system of referral and recommendation—but to actively promote your own practice was genuinely frowned upon.

Then, in 1977, the Supreme Court passed a landmark decision permitting professionals to advertise their services. Suddenly, those in the professional services industry found themselves in the open marketplace—able to employ all the marketing strategies that were previously withheld from them.

Of course, tax practitioners are trained to fulfill their tax responsibilities in a technically competent and satisfying way; they are not necessarily familiar with the complex field of contemporary marketing and promotion. Nevertheless, marketing—over the years—has certainly gained in prominence as an essential component in the development of a successful tax practice. Therefore, it is important that you become at least somewhat familiar with the field of advertising and marketing. Hopefully, the material gathered in this chapter will be helpful in at least introducing you to some basic marketing ideas, while providing you with a number of practical suggestions for handling this vital aspect of your practice.

It has been years since the Supreme Court delivered its decision in regard to professionals advertising, but there is still some residual tension between the more traditional practitioners who favor 'building up a practice from within'—through referral, and the 'new school' whose counter argument runs: Business through referral is fine, but a new practice has no clients to begin with; so how is referral to ever take place? They go on to reply that the marketplace is radically different today, then it was 20, 30, and 40 years ago; competition is keen, and the circumstances demand new and different strategies to attract clientele.

It could also very well be that everyone in this debate is right. You just have to bear in mind that a developing tax practice goes through phases, or stages of growth, and each phase calls for its own type of market strategy.

To open a practice...to get off the ground...to make the public aware of who and where you are—is a little like launching a rocket into space—and it calls for a certain concentrated boost of energy. At this point you may have to rely more heavily on outside advertising and marketing. But once your practice is active and ongoing, then you may rely less on outside advertising and emphasize 'marketing from within', i.e.—improving and developing client relations and working through referrals.

And then again—if you should move, take on a partner, or expand into new or specialized areas—you may need to run a new advertising campaign. It all depends on what you want to do with your practice at any given point in time.

In any case, everyone agrees that the fanciest and most sophisticated advertising campaign cannot guarantee you success; because you are the central ingredient to your own success. A good ad may lead to initial contacts and new clients, but only you can keep those clients coming back season after season—by virtue of the quality service you provide. And then it follows that a growing and satisfied client base will gladly generate more business through referral.

One last point: Even when the Supreme Court handed down its decision allowing professional services to advertise; it was with a sense of discretion and a clear set of guidelines. In so many areas, much of what passes for advertising is downright gaudy, demeaning, and misleading; and you don't want any of these features to reflect upon your own professionalism and integrity. So remember—when you do advertise—don't make false or misleading claims. Be tasteful and tactful, rather than flamboyant; be honest and straightforward, without leaning toward exaggeration and sensationalism; and emphasize your own merits rather than disparage your competitors. In this way, your marketing efforts will be a sincere and reliable extension of the quality tax service you have to offer.

Marketing: An Introduction

Marketing is a broad, all encompassing term that refers to almost anything you professionally say, write, or do, and that relates to you and your tax practice. It includes all the various ways in which you, as a tax professional, communicate with and appear to—the public around you: the way you look, dress, and talk; the appearance of your office; the letters you write; the ads, fliers, and brochures you print; the organizations you support and belong to—however you express yourself in public can relate to the concept of marketing, in the broadest sense of the word.

As you can see, marketing is a complex, yet fundamental part of your practice, and will demand your ongoing attention. It's one thing to be an accomplished and qualified tax practitioner; but it's quite another to communicate the message to the public. And that is what this chapter is all about: effectively communicating your professional image and message to the public.

Since marketing is such a multi-faceted subject, we will discuss the subject by breaking it down in the following way:

- Market research: identifying and learning about your target market;
- Market planning and strategy: where we discuss the different forms of advertising;
- Publicity;
- Promotions;
- Public relations.

Market Research

Your overall goal in marketing is to effectively communicate with prospective clients in order to elicit a positive response. And market research is a vital preparatory step that aims to help you attain this goal. It involves gathering information about prospective—and current—clients, so that you can make more intelligent marketing decisions. The more you know about your clients—who they are, where they are, and what they do; what they read and what they need—the better you can structure your advertising message, publicity campaign, and other information sharing programs.

Your Target Market

Your first objective, then, is to identify the group (or groups) of clients you feel best qualified to serve. Not 'everyone' out there is a potential client of yours. There is no service in the world that everyone partakes of in the same manner, or at the same time. Some members of the purchasing public are either too young, too old, too rich, or too poor, and some already have a tax consultant, and so on. At the same time, you cannot possibly be everything to everybody; so who are you best qualified to serve? You have to be aware of your specific skills, strengths, and limitations, and thereby recognize your potential market niche—somehow classify the public you are in a position to serve—

then tailor your marketing effort in that direction.

To help you locate prospective clients, some measure of market research is necessary; but don't worry—this is not a complicated procedure.

In a previous chapter we discussed the value of demographic studies in relation to office location, and this same body of information can help you locate potential clients as well. In fact, most—if not all—the information you need is readily and locally available to you.

Remember: demographic information can identify individuals in various ways—by age, gender, education, income level, residence, or occupation. And by applying demographic findings you are able to break the overall market down into subgroups based on these specific characteristics. You can address yourself exclusively to a specific geographic locale, to young singles, young families, or senior citizens. You might want to focus on a particular age, or certain professions, or owners of small businesses—or combinations of the above. Over a period of time, you might test various messages on different populations, and see where you strike the most responsive chord.

You can find the demographic information you need at the following facilities:

- Your local library;
- University business departments;
- The Department of Commerce;
- The Chamber of Commerce;
- The Census Bureau;
- Small Business Institutes;
- Utility Companies;
- Regional and local planning commissions;
- Periodicals such as The American Demographic Magazine or The Sales and Marketing Magazine;
- You can also discuss these matters with local public officials, and people in the business world.

Once you have gathered some information and have identified your target market (or markets), you are ready to devise a message for that segment of the population.

Your Marketing Message: a good place to begin—

At first glance the following exercise may seem simplistic, but it is well worth your time...

As best you can—in writing—clearly describe exactly what you do professionally. Briefly state the services you render: you can list them, mention specialties, even elaborate on each point if you want. You might even draw up several versions of this exercise: a concise statement as well as a more drawn out explanation. But however you formulate the sheet, when you finish you should have a precise description of the services you provide as a tax professional.

Now on another sheet, list and describe all the benefits and advantages you offer your clients. Formulate all that you can do for them: Do you save them time? money? energy? bother? confusion? worry? Do you provide them with a sense of confidence? security? clarity? peace of mind?

After drawing up these sheets you will have the two essential sides of your 'marketing message': One sheet addressing—'What I can do'; and the other sheet—'What I can do for you'. And this is basically the heart of your message.

In one form or another, this is the message you want to communicate to the public. You can always modify it, shorten it, lengthen it, highlight certain features over others—all depending on the purpose and context in which it appears—but the message remains basically the same for you to draw upon and work with as the need arises. In a greatly reduced form it may appear on a business card or flier. In a more elaborate form it may fit into an article or brochure. But once you have taken the initial time to express and formulate this information, you will always have it to turn to. The tone and content as it reappears in various formats will also lend a consistency and professionalism to your practice, to the literature you produce, and will provide clients and prospective clients with a clarity and confidence about your practice as well.

Targeting your market and formulating your message are two fundamental and important considerations in launching a successful marketing campaign.

Learn From the Experience of Others:

Another worthwhile source of information to help you in market research is to read the results of articles and studies that relate to your field. From time to time articles appear that bear directly upon tax professionals and the public they serve: some inform readers of what to look for in a tax professional; others report the experiences of practitioners in such areas as marketing or public relations. By reviewing such articles you are in a position to benefit from the research, findings, and experiences of others—by applying the knowledge gained to your own practice.

For example, one financial journal recently instructed readers in how to shop for a tax professional. Just briefly the article stressed the following points:

Look for—

1. Straight talk about fees... spelling out clearly how fees are to be calculated.
2. Easy access... Calls should be answered promptly, in or out of tax season.
3. Clear explanations... of everything the tax professional is doing on the return, including deductions and other items.
4. Education... time spent with a tax professional should give the client a better understanding of how tax law affects him; the client should then be more aware of what records to keep and procedures to follow.
5. Year round tax tips... by November a tax professional should contact clients and alert them to last minute ways of cutting tax bills.
6. Long range tax planning advice... a good tax professional should provide clients with financial tips that go beyond the current tax return.
7. Audit-aid... a tax professional should be willing to advise and accompany clients on an IRS audit.

In another, related report readers were told to apply the following criteria when hiring a tax practitioner.

1. Does the professional answer your questions with ease and with apparent technical competence?
2. Is the professional up to date on recent changes in tax laws, recent IRS regulatory changes, and current court decisions?
3. Does the professional provide you with a comprehensive analysis of your tax situation in clear and understandable language?
4. Does the professional spend time discussing legitimate methods to reduce tax liabilities?
5. Does the professional provide you with practical audit planning suggestions?
6. Does the professional keep appointments on time? Is he/she accessible and prompt in returning telephone calls?
7. Does the professional provide you with a clear understanding of fees and costs, up front? Is the price reasonable?
8. Will he/she represent you at any audits, and if so, at what cost?
9. Are the two of you compatible?

One firm of tax professionals wanted to know specifically how they could improve the firm's services, so they hired an agency to objectively survey their clients in person—through interviews and written reports. In such an undertaking, not only the agency, but the clients too, were compensated for their time and services. And around the lessons learned from this particular survey were the following:

1. It was discovered that in this firm the personality of the principal partner was the most important reason for clients choosing this firm over others—all other features of a firm's competency being equal.
2. Many clients felt it detrimental that the firm was losing its 'personal touch' since the tax preparation procedure had become increasingly computerized. In fact, on account of this, some clients were seriously contemplating a switch from one firm to another.
3. The firm also learned how important is was for each client to clearly understand the basis and breakdown of the practitioner's fees.

Guidelines such as the ones listed in the 3 examples above can be helpful in several ways:

- They can serve as a standard by which you can examine and assess your own performance level in your practice. Just ask yourself: To what extent do you fulfill these expectations?
- By being aware of such concerns you can improve the quality of your interactions with clients. By being sensitive to and openly discussing these matters you create a climate of confidence and trustworthiness in the practitioner/client relationship.
- You can also emphasize these concerns in your marketing efforts. Knowing that these are important issues to your public, you can address yourself to them in your advertising.

As a rule, then, to know what your prospective clients are interested in and concerned about is an essential part of market research and planning. If you can satisfy these needs and allay these concerns, then you may be just what the prospective client is looking for.

How Feedback From Your Clients can Strengthen Your Practice:

Until now our discussion of market research has centered on how to effectively utilize 'outside' data... findings you can gather from beyond your office walls. But you can also gather helpful information from your own clients. It would indeed be advantageous for you to know what perceptions your clients have of your practice: what are their needs, expectations, and changing concerns; in what ways might you improve the services you offer them; or what additional services would they like you to provide. Getting a handle on such information is certainly worth your while, and there are several ways in which you can go about gathering it. Of course, bear in mind that you must always be tactful and considerate; you are dealing with your own clients and you must be sensitive to the manner in which you elicit such information from them.

1. Depending upon your rapport with any particular client—if its an open, established relationship, based on shared confidences—you may find that you can get the answers you need by speaking to the client directly. By asking brief, relevant questions, the client will share with you the feedback you are looking for.

2. If you think you would get a better, more objective response if the client answered in writing—and even answered anonymously—then you could send out a brief cover letter and questionnaire. The letter could open with the following remark...

    Dear Mr. Smith,

    Client satisfaction is one of the most important services I can offer you. I am always interested in improving the quality of my practice, and would like you to help me by answering the following brief questions.

    Thank you for your time and information, and I hope to be of service to you in the future.

    Respectfully,

    Then construct a list of important, specific questions highlighting key features of your tax practice procedure. To make things easier on your client you might follow each question with a rating scheme of 1...5; moving from poor, fair, good, to very good, and then excellent. After each question you could still leave space for the client to enter any personal remarks and suggestions if he wishes to.

    The following is a list of possible starter questions; you can add, subtract, or modify—tailoring the questions to meet your own needs.

    A. How quickly do we respond to your inquiries?

    B. How accessible are we to you?

    C. How courteous and helpful are we to you?

    D. How would you rate the quality of our service to you?

    E. Would you recommend us to others? If yes, why? And if no, why not?

    The more specific the question is, the more applicable it is to various aspects of your tax practice, the more valuable the answers will be to you. The point is to try and get an accurate reading on pertinent and specific aspects of your practice and quality of service.

3. You can also gather client information without asking your clients anything. Simply write down your own personal observations, and study more closely the data you have regarding your client, in order to get a clearer profile and better

understanding of the types of clients you service best. Then, based on your findings, you can:

A. Devise methods to better service the clients you have, and -

B. Devise marketing strategies that appeal to potential clients who are similar—in significant ways—to those you presently serve.

A successful tax practice is not a static affair: it is—instead—a dynamic, evolving enterprise, and some measure of market research is always called for just to stay in touch with the needs of an ever changing marketplace.

The more inquisitive and methodical you are—to gather pertinent client data for the sake of your practice—the better prepared and informed you will be to handle both your clients' needs, and those of your growing practice.

The Market Media: Planning & Strategy

Before discussing the various advertising media 'per se', we will first briefly describe the following essential marketing tools: your professional name, your logo, business card, stationery, and brochure. These are important marketing items: they describe and represent you and transmit to the public an image of your practice. When you mail, post, or distribute any of these items, it is like leaving a bit of your professional self behind. They serve to identify you as a tax professional; and though they are an essential part of your overall 'marketing kit', they are not really examples of advertising—in the strict sense of the word.

What's In A Name: Selecting an Appropriate Name for Your Practice...

Choosing an appropriate name for your practice is an important—but pleasant task. You want to select a name that you like, that is easy to say and remember, that relates to your practice, and that has a professional ring to it. So before you decide upon a name, take the following considerations into account:

- Try to keep the name straightforward and descriptive. If possible, the name should somehow indicate what you do rather than raise questions or doubts about your services. For example: Adams Tax Assistance rather than Adams & Associates.

- Try to keep the name short, specific, and pronounceable. Long, cumbersome names are difficult to deal with and easily forgettable. If your own name fits the formula, then use it. But if your own name is too long and difficult to pronounce, then think of something else.

- Try to think of a name that is distinctive and memorable; and be sure to check if the same name is currently in use, to avoid later business and legal complications.

- Try to avoid clever and humorous names. For example: 'Give Me A Break Tax Consultants' may seem cute at first but you have to remember... some may not think it funny, and others may simply write you off as unbusinesslike. In any case, when it comes to naming a professional practice, humor is a questionable marketing tool and may not serve you well.
- Also, try to stay away from trite, overused terms, or superlative, grandiose sounding words like 'Ace Tax Consultants' or 'Universal Tax Service'... these are overworked terms and tend to have a hollow ring to them.
- Finally, get some feedback from people you trust before settling on a professional name. You want to be comfortable with the name you choose, and you want it to project a professional image to the public.

In addition to a name, a short, effective slogan can enhance your image and make your name and practice more memorable and appealing to the public. If you do use a slogan, just be sure it suits your image: keep it short and to the point, and stay away from silly, exaggerated remarks.

Logos

To further complement your business identity you may want to select a logo to enhance or accompany your name. Choosing a distinctive typeface or some sort of type configuration may suffice for a logo; or you can have someone design an appropriate graphic symbol to accompany your name. A good logo can be eye-catching, and get people to notice and remember your practice.

If you do select a logo, be sure to use it extensively: on signs, business cards, ads, stationery, bills, letterheads, brochures, press releases—anything that comes out of your office. It will help you achieve quick, easy, and extensive recognition.

Letterheads and Envelopes

Your letterhead refers to your business stationery; and it should be used for all your business correspondence. Having such stationery lends credibility and professionalism to your practice; and if possible, it should contain your logo, and certainly your business address and phone number. A well designed letterhead should also serve as the basis for all your press releases and other informational literature.

Business Cards

Do not underestimate the value of a good business card; it can be a very effective, creative, and versatile marketing tool; so be sure to have cards printed as soon as possible. Then—distribute them freely: attach them to your mailings, press releases, letters, and direct mail campaigns; leave them at appropriate public places, with CPAs and bookkeeping services, or attached to bulletin boards; and whenever you deal with another business or individual, be sure to politely give him your card.

A few other brief tips may prove helpful:

- Before you print your own, examine other cards and see what attracts you to some and not to others: Is the primary information legible? Does it stand out? Is the logo or name a memorable one? Is it clear from the card what service is being offered?
- It may be worth your while to print on both sides of the card and embellish the second side with added information, giving a recipient more reason to hold on to the card. This second side could be used for anything from a chart of weights and measures, a conversion scale, or a tipping chart, to a short inspirational message, tax tips, or even a brief rundown of the services you render.

Brochures

Brochures are an effective, relatively inexpensive marketing tool used to inform the public of your services. It can deliver more detailed information about your qualifications, expertise, and the services you render—more than can appear in an ad or even a flier. It can be an effective mailer on its own; it can accompany a press release; it can be available or distributed at public gatherings, presentations, or simply left at offices, schools, and other public facilities.

A good brochure should clearly identify you, your tax practice, then list your areas of expertise, describe the extent of your services and what benefits you can provide for your clients; and if there's room—it can include a brief biography listing qualifications, awards, honors, and interests.

A good brochure need not be expensive or thought of as a complicated, extravagant project. The point is to explain who you are, what you do, and what you have to offer—attractively, but with an economy of space and words. And for this, a single sheet of standard sized paper—8-1/2" x 11" folded in thirds, or 8-1/2" x 14" folded into thirds or fourths—will suffice. You can print on both sides and even leave one panel free for mailing purposes. Of course, brochures can get lengthier and fancier, but if you select an appealing typeface along with complimentary graphic material, you can certainly come up with a well designed—inexpensive, yet impressive—piece of work. You might even incorporate some tax tips or other worthwhile information into your brochure; this will give the public added reason to hold on to the brochure for future reference.

Bulletins and Newsletters

It may not be an immediate consideration, but when the time is right, a periodic client bulletin or newsletter can be a most effective way of maintaining contact with clients, or establishing contact with prospective clients. It can serve the dual purpose of informing the public of vital tax related news, while promoting your practice in a meaningful, constructive way.

Whether you call it a 'bulletin', a 'newsbrief', or a 'newsletter'—it doesn't really matter—the terms are basically interchangeable, and the format is primarily the same. To begin with keep your style and content simple; one page 81/2" x 11" or 81/2" x 14", printed on two sides. Use your logo and put a masthead across the top to give your publication a professional look. Be sure to incorporate information that your public will find useful. Material for such a bulletin can include news briefs, excerpts from pertinent articles, tax planning ideas, new regulations and requirements, or other informative features. For ideas you can draw upon any number of resources: weekly tax services, IRS bulletins, accounting publications, business journals, questions raised by your clients, and your own ongoing office experience. And you can always put in short fillers: personal anecdotes, interoffice news and changes, tidbits, sayings, and humorous asides. You can start by coming out 4-6 times a year, and if you do use material from other periodicals, be sure to quote the sources in your publication.

If anything, newsletters are usually lengthier, and more elaborate than bulletins, yet they draw upon a similar pool of material, and at a certain point of professional development, they are an excellent way of broadening and developing client relations and services.

Newsletters tend to be an expensive and time consuming venture, but you can also purchase one of the commercial newsletters that are available, affix you own logo or letterhead to the top of it, and send that out to your clients and other contacts.

But whether you decide to work with a brief bulletin or a lengthier newsletter, both are excellent marketing tools: they show concern on your part, they are informative, they serve as a communication link to regular and even occasional clients, and the public becomes more aware of what you are capable of doing. Finally, such communiques can generate a lot of response, with people calling, commenting, and asking questions. You may even want to have a reader response column where you answer interesting inquiries from those who write in.

Advertising

At this point, assuming that you have some idea of a target market, and a message in mind, we will deal with the subject of advertising.

Advertising is basically any paid communication you devise and transmit via any of the media available to you. We often think of the 'media' as referring exclusively to such mass media as radio and TV, but it encompasses all of the print media as well—newspapers, magazines, posters, billboards, mass transit ads, hanging signs, etc.

Each of the media has its own distinct advantages and disadvantages, and you will have to weigh your options to decide which medium or 'media mix' (where you can make use of more than one at a time) you wish to employ.

We will now briefly discuss them individually.

Electronic Media—Radio and TV Advertising

From a financial perspective radio ads may be an affordable type of advertising if you use local stations, but there are other factors to consider. You have to remember that people usually listen to radio when they are doing something else... dressing, driving, washing dishes, and so on. Therefore the all important specifics of your message may get lost. Those listening may not be able to write down or remember your number. Also, a lot of listeners tune out ads when they come on, or switch stations. However, if you do use radio, remember that the key is repetition, and timing is very important as well. Place your radio ads at the appropriate time of the year for tax considerations, i.e. prior to the peak of the season. Another point to consider if you use radio is that to a certain extent you can select your audience, insofar as certain types of people listen to certain types of programming on specific stations.

Using television is probably the most expensive form of advertising. It reaches the most amount of people at any time, and it is visually impressive, but remember too: the message is fleeting; it may be wasted on many who are not really part of your target market; and yours may not be the type of message that needs TV visual accompaniment to begin with. Also, the point made above regarding radio applies here as well—that people either tune out or switch channels when ads come on.

Considering the pros and cons of the electronic media, and judging by the experience of most tax practitioners in the past, it would seem that your advertising dollar would be spent to much better advantage in the print media—which is the next subject of discussion.

Print Media—Newspapers

Newspapers primarily serve local markets and concentrated populations. Of course, 'local markets' can mean a small town or a large metropolitan area, but whichever the case may be, your newspaper ad will basically reach out to the locals serviced by the particular paper.

Newspapers are immediate and practical sources of information, and most Americans consult with them daily; so advertising in newspapers at least assures you of a lot of exposure. However, studies indicate and most advertising managers agree that newspaper ads require repetition to be genuinely effective, so choose smaller ads that appear more regularly than a large ad that appears less frequently. Remember too, that newspaper ads have a short life span—about one day—after which they are usually discarded. There are also a lot of other ads or features that your ad must compete with on any given page, so your ad may be overlooked or lost in the confusion.

To a certain extent, when you advertise in newspapers, you can select your audience by placing your ad in a specific section of the paper or in a section that appears on a specific day of the week (sports page, business section, a 'home' or 'family' page, and so forth). You thereby increase your chances of being noticed by the market you aim to address. Nevertheless, to a great extent, much of a newspaper's circulation is 'waste'—as far as your ad is concerned—simply because a substantial number of readers will not pick up on it: they will either overlook it or disregard it either because they don't need the service, or they live too far away, or for some other reason. Therefore, advertising in smaller papers—neighborhood papers, suburban papers, special interest papers, even weeklies or monthlies—may prove more rewarding than advertising in large metropolitan papers.

Display ads—the larger ads that appear on pages that carry other features or news articles—can be quite costly, whereas classified ads are much cheaper and therefore easier to repeat. And although classified ads are not seen by as many readers as are display ads, those who do read the classifieds are more likely to be in the market for something specific.

Print Media—Direct Mail

Direct mail is advertising literature that is sent through the postal system, and for the tax professional it has several distinct advantages over other print media advertising channels. If carefully planned and constructed, and properly executed, direct mail can be a most effective way of reaching new clients.

People oftentimes—and erroneously—refer to direct mail advertising as 'junk mail'; but it is never junk mail unless it is the wrong message sent to the wrong person, or unless the piece is so poorly conceived and assembled that it is not worth paying attention to. Otherwise—if the message is clear and creatively executed, if the message speaks to or answers a definite need, and if it comes to the right person at the right time—then it is most definitely not junk mail, but an effective and welcome promotional tool.

Direct mail has the following advantages:

1. For one thing, in a broadcast or print ad, you have to be very selective and mention only brief highlights relating to your tax service. But in direct mail you can elaborate and tell as much or as little of your story as you wish, you can modify your message and speak more directly to specific needs, and you

can describe various problems and solutions. You also have greater flexibility as the nature of the material you relate—both in format and in content. Your mailing piece may be little more than a postcard, or it can be a fold-out sheet of paper, or it can even serve as a brochure, in fact, you can design something that can function as a brochure as well as a direct mail piece.

2. You can also target your direct mail advertising with great precision, because you are in control of where your direct mail piece is sent.

   You can develop your own mailing list from current clientele, referrals, family and friends, or you can purchase or rent ready made lists that virtually pinpoint almost any segment of the consumer market that you wish.

   You can also use the option of 'occupant' or 'resident' mail if you simply want to blanket a certain geographic area. Your local post office can supply you with the necessary details.

3. With direct mail you can repeat or follow up on your message whenever you want and time your mailings to more precisely meet your needs. You can also be more precise in measuring the responses to your mailings and calculate their effectiveness.

   The bottom line to all of the above is that you have more 'freedom' and more 'control' over all aspects of advertising in a direct mail promotion. It's flexible and selective at the same time.

4. Another worthwhile feature of direct mail is that it is very effective when used in conjunction with another medium—especially the telephone.

   When one medium serves to complement or reinforce another it is commonly referred to as a 'media mix'. And toward this end, a direct mail promotion piece fits in well with a telemarketing campaign as part of a single marketing strategy. The piece received in the mail usually paves the way for a follow up call within a week.

   One final thought: Experts in the area of direct mail caution those who are just beginning...to make small mailings at first and expand from there. You may even want to sample certain areas, gauge your response, and then move to another area, until you find where your response is best.

   Remember, too, that major mailers consider a 1% response to be quite good; they would be ecstatic about a response of 3%—4%. So be realistic in your mailing... and be persistent!

Direct Mail—Lists:

Since direct mail is the most personalized and targeted of all the print media, a well planned mailing list lies at the heart of every successful direct mail campaign. Each mailing list is distinguished by its own unique features, and you will have to determine what type of characteristics you will want to use:

- You can develop your own list beginning with friends, neighbors, relatives and acquaintances;
- You can compile a lost of potential clients on your own. For example, if you want to mail to certain businesses or professions, you can gather names from the local yellow pages.
- If you think certain neighborhoods are good prospects, you can get voter registration lists for a specific area.
- Otherwise, you can buy lists or rent them from list rental firms.

Direct Mail—Timing:

Many products and services are closely related to a specific season of the year; therefore you will find soft drinks and air conditioners getting a lot of media attention in the summer, while hot chocolate and snow plows get more in the winter.

Similarly, as a tax professional, you should concentrate most of your advertising efforts into those months leading up to and including the tax season. It's just common sense that most people are more receptive to what you have to offer at the appropriate time of year. This holds true for all your media exposure, and is most certainly the case when it comes to such a concerted effort as an extensive direct mail campaign.

Print Media—Directories

A directory is not usually an active, persuasive advertising medium, but rather—a reference medium. People usually turn to a directory when they are looking for a particular product or service. Among the advantages of placing an ad in a directory is that it tends to have a long reader life; and it is a practical, popular directory (such as the Yellow Pages), then usage can be quite high and consistent.

You may want to settle for a regular directory listing or a bold print listing; but you may also want to consider a display ad—business card size or ever larger; because if you place a distinctive, attractive, practical ad in the Yellow Pages—or some other directory—then those individuals actively looking for a tax practitioner may just decide to call you first.

It is also important to be listed in as many directories as possible—Yellow Pages, community business listings, neighborhood directories—or any place where potential clients may come across your services.

Tips For Writing An Effective Ad

1. Select a brief headline or lead-in that captures the reader's attention: offer a benefit or ask a potent question. Many readers don't get past a headline, so this part of the ad must have direct appeal.

2. In the body-copy of your ad you can elaborate and be more specific: describe your services and what you can do for the reader.

3. Speak in positive terms, use everyday language, short sentences, and write sincerely. Be personable and direct: speak to the reader using singular "you".

4. Be believable: don't make outrageous, sensational claims. Don't use subjective phrases or opinions like, 'we are the best' or 'we are the only...'. Instead, be straightforward and point out the very real advantages and benefits of your practice.

5. Use an attractive layout and design. If there is room for graphics to accompany your text, it would be helpful.

6. Be sure to include the name of your tax practice, address, directions if necessary, telephone number, available hours, and your logo if you have one.

7. Try to have someone else proofread your copy. Whether it's for an ad, a brochure, or an article very often you are so close to the project that you fail to notice even obvious mistakes. Someone whose standpoint is more removed—and therefore more objective—is in a better position to catch these errors.

8. A related point: As you get involved with writing ads, you will very likely get involved with typesetters (unless you do your own typesetting and graphics) and printers as well. Good typesetters and printers will take the time to work with you and answer your questions, so try to find those who will prove to be helpful companions in your growing enterprise.

   It's a good idea, anyway, to get quotes from more than one typesetter and printer. Get personal recommendations, and in every office you visit, be sure to look at a sampling of the work they produce.

9. Also bear in mind that you are selling a service rather than a product.

   Oftentimes, service benefits are more intangible than product benefits, and it's harder for the public to 'picture what they are getting'. Therefore, when you formulate and structure any of your print information, try to be specific, and make what you have to offer as clear, concrete, and graphic as you can.

Keeping Track

Another important—and ongoing—part of market research is to track the results of your advertising. You need to know which media, and which messages, and which times of the week, month, and year are working best for you, and which marketing efforts are just a waste of time and money. For example, you might test various ideas on different segments of the population and discover that a certain message does not work well with one group but does for another. So you have to evaluate your results and refine your strategies in order to get an ever better return on your advertising investment.

By simply asking clients, one successful practitioner was surprised to learn that the key selling point in her advertising was the statement that 'the return will be prepared in the privacy of your own home or office'.

Over a period of time then—by keeping track—you will get an ever clearer indication of what marketing techniques are working best and who is responding to them.

Another interesting result of your research may be that you will uncover a 'market gap'; you may come upon a segment of the population in need of your service that you never even considered before. It sometimes happens that the needs of the larger market will help determine the shape and direction of your own practice—and it may mean moving in unexpected directions rather than according to your plan. But as an up-and-coming tax professional, you will always have to be on the lookout for such marketing opportunities and new ways of reaching clients.

What follows is the personal account of one practitioner's experience with advertising a new practice. The article appeared in a leading national accounting periodical, and the results of the author's findings appear here in abbreviated form. And although this is far from being 'the last word' on the subject, the author's insights may prove helpful to you...

Phase 1:

1. The author relates that he first clearly defined for himself the 'services' he was selling and the market he was targeting. He chose to emphasize tax service, financial planning, and write-up work.
2. The initial goal of his ad campaign was to gain exposure and name recognition. He placed ads in several local newspapers (in late November to get pre-tax season exposure) to announce the opening of his practice (listing available services, hours, evening availability, phone number, and office location).
3. The 'new office' ads led to many calls and several clients.

Phase 2:

1. From late December to April 1, the author aimed at obtaining a large number of tax preparation clients. When he spoke to prospects he learned that his best results came from ads in local suburban papers, ads in a local specialty business

paper brought in average results, and spot ads on the radio brought in the poorest response.

2. The author has an enthusiasm for horses and raising horses, so he put an ad in a paper that caters to fellow enthusiasts, and this ad generated an excellent response: out of 20 phone calls, 19 resulted in new clients.

3. Ads in weekly suburban papers paid off well and new clients responded especially well to the availability of Saturday and evening appointments.

Phase 3:

1. The author's third goal in advertising was to obtain year round engagements with clients, and to this end he found that display ads in the yellow pages were very helpful.

2. The author continues to monitor new clients to determine which ads work best, and he seeks further exposure by teaching tax, money-management, investment and other finance-related subjects, public speaking, and attending functions that reflect areas of personal interest.

3. The author concludes that his ad campaign was successful: he developed a substantial client base in his first year of practice; he learned where to spend his advertising dollar; and he proved to himself that a successful new practice can be established without an existing client base—with the help of an appropriate advertising campaign.

Educating the Public- A Type of Market Strategy

It's amazing how little most Americans know about the content and structure of tax law. And then again, it's understandable too: tax law is complex, undergoing constant revision, and given society's hectic pace—most people just don't have the time or patience to keep up with it.

Consequently, many of your potential clients may not even be aware of how a tax professional can be of help to them They may have misconceived notions of what a tax professional does, or who he services; or they may not have any notions at all. And being unaware of the benefits, they may never avail themselves of the services.

Therefore, you may find it advantageous to make 'educating the public' an integral part of your advertising campaign. Make it a point to inform prospective clients in precisely what way you can help them save time, worry and money. You can incorporate such information almost anywhere: in a brochure, in a brief message on the back of a business card, on a flier, in and ad, as an article, in a column of tax tips in an office bulletin, or as part of a direct mail campaign. And perhaps most important of all: educate clients in the course of your daily interactions with them. Coach them in how they can help their own financial situation along; show them how they can prepare themselves to meet with you; make them aware of up and coming tax law changes in tax policies that may affect them in one way or another. You will find that the more informative you are—the more trusted you will become. Your extended and personalized service will be greatly appreciated, and you will have acquired a client for life.

Educating the public, then, can prove to be a worthwhile part of your marketing plan; not only are you a source of convenience and professionalism, but a source of reliable information as well.

Neighborhood Networking

Advertising studies bear out that those who live and work in a specific area respond to neighborhood marketing conducted by neighborhood business people. This is another good reason to plant yourself in a carefully selected location.

With this in mind, think in terms of reaching further out—by phone, card, brochure, or in-person visit—to facilities that operate in your neighborhood:

- Neighborhood businesses—their customers and clients;
- Neighborhood employers—and their employees;
- Neighborhood associations and organizations—and their members;
- Neighborhood educational institutions—and their staff and students;
- Neighborhood non-profit groups—and their supporters;
- Some of these groups may publish their own newsletters or bulletins. Put an ad in or offer an article to these publications;

And, finally, don't forget the importance of your daily interactions with people on the street. Your conduct in such a setting can go a long way to promote your professional image. Be courteous, patient, of good cheer; and the neighborhood's opinion of you will grow in a positive way. Furthermore, if you have bulletin boards or window space open to public view, allow community groups and organizations to post their notices and events there. This too will enhance your neighborhood image while—at the same time—attracting traffic to your location.

Publicity

Basically, any mention of your practice in the media—that you pay for—can be classified as advertising, while any mention of your practice that you do not pay for can be termed publicity. The information may be solicited or unsolicited, but when the media makes use of it, you get the publicity; so it amount to free communication. And although publicity can take place in any of the mass media, the most common form—and the one we will elaborate upon—is the written press release.

A press release is a primary means of relaying newsworthy information about you and your practice to the print media at large. A publicity press release can inform, explain, or educate; you just have to pick out a noteworthy point to expound upon. Here are just a few suggestions for features, stories, or articles: grand opening events; expansion of services; new partnerships; outstanding achievements or honors earned; human interest stories or public service activities. You can write about a unique aspect of your practice, and if you can't think of one—then create one! For example, provide a free consultation day or learning seminar, and then write up the event—both before and after—as a press release. You could also provide a comment or interpretation of tax law that effects a lot of people; or propose a solution to a common tax problem, and submit the article as a press release.

Remember too: don't overlook or underestimate neighborhood or local press coverage, just because it isn't the International Herald Tribune. Be realistic and practical: the smaller, more community oriented resources of publicity—such as weekly and suburban newspapers, shopping bulletins, church bulletins, public bulletin boards, newsletters, periodicals from special interest groups, clubs and service organizations—these channels are usually more willing to use your material, and these resources are generally more in touch with your potential target market as well.

Of course, if the opportunity arises whereby you can appear on TV or be interviewed on the radio, don't pass up the chance. But at the same time, don't overlook the opportunities that are close at hand.

Here are some suggestions to help you write your own press release:

1. First, identify a newsworthy event or piece of information (use suggestions above), and then be sure to write your article from an objective point of view, in the third person—as if you are merely reporting the event or information. Don't write in a way that appears as though you are 'patting yourself on the back'.

2. Select appropriate publications for your press release. Consider the various publications available to you, and then select the ones you think are most suitable—and likely to print your pieces. In fact, study the releases that appear in those periodicals; familiarize yourself with their style and content. You might clip and keep a file of releases for future reference. This way you will be better able to tailor your own writing for publications.

3. When you write your article, keep the following points in mind. a) Write a short, attention grabbing headline; and a first sentence that sparks further interest—to hook the reader. b) Write the more essential and more interesting

information at the beginning of your article, and put the less important information at the end of the piece. c) Do not write long sentences. Keep the language simple and straightforward, and limit your article to somewhere between one and two typed pages. d) Your press release should have a neat, crisp, businesslike look: typed, easy to read, double spaced, with even margins all around the page; and submitted on your own letterhead stationery.

4. In addition to your name, firm name, telephone number, and person to contact for further information, your release should also include a) A release date. Most press releases carry the line FOR IMMEDIATE RELEASE which informs the media that the article can be used right away or at the editor's earliest convenience. b) Suggest a headline that briefly summarizes the content of your press release in an exciting way c) Include a copy of your brochure or business card together with your mailing.

5. Be sure to send a gracious thank-you letter to anyone who uses or prints your release; it's common courtesy to do so and such acknowledgment may pave the way for the publication of other articles and letters.

6. Most of all, be persistent in your efforts and they will finally pay off. In your contact with editors and program directors, remember that they are not looking to give you free publicity, they are looking to inform and entertain their readers, listeners, and viewers. Shape your release with that viewpoint in mind. If you write in an obviously self-serving way, you will not get the publicity you're after. But if you simply persist—bearing in mind that not everyone will print your releases—you will see that your efforts will eventually serve you well.

7. Remember, also, to be resourceful: If you wrote a press release, or an article, and it found its way into print, then turn it into reprints that can serve as handouts or mailers. The reprint can be attractively assembled with your logo or letterhead across the top. You can then send it to clients or prospective clients signed at the bottom—"With compliments from..."—followed by your name.

   Similarly, try turning your press release into a more full bodied article; or turn your article into a speech, or your speech back into an article or press release. In other words, use your imagination, and try to get as much mileage or publicity out of your efforts as you possibly can.

   Furthermore, keep a running record, or scrapbook, of all your press releases, it will prove a worthwhile effort for your own self and will impress clients favorably as well.

Aside from press releases, here are some other opportunities for publicity:

1. Write small, filler articles for newspapers. These can be anecdotes, tax-tips, about 50-80 words in length, and you can send several at one time.

2. Write letters to the editor. Provide a timely reply to a subject you read about in

the newspaper or some topical issue. It may not even be related to your tax practice, though it helps if it is. Such editorial replies usually run between 200-250 words.

When you write such letters, remain focused on the subject at hand. Rely on logical, straightforward arguments, and well documented facts and figures; and be sure to sign off with your name, firm name, address, title, and telephone number. If you have any further questions about such letters, study some samples from the periodicals you wish to address, and model your own replies after those.

3. If you are appearing somewhere—to speak, give a seminar, or to offer advice and answer questions—be sure to notify the media beforehand, they usually allot some measure of time and space for public service messages. Contact local newspapers and radio stations and ask what their procedure is for printing or airing such information, it's a valuable source of publicity for you.

The most effective form of publicity, after all is said and done, is word of mouth recommendations from your own clients. Satisfied clients who tell their relatives, friends, and work associates about your practice, can establish and further your credibility in a way that no other planned strategy can.

Especially when it comes to the realm of personal services, people simply recommend to others those professionals they have come to know and trust. Therefore, the ultimate focus in a successful tax practice—is simply you. To generate a loyal clientele, you have to make sure you provide the kind of quality service clients will want to tell others about. And that's the best publicity of all.

Promotional Events

Promotional events are special occasions designed to increase contacts, exposure, and general awareness of the services you provide. Such occasions may include the opening of a new office, or a change of address; they can promote a unique and timely service; they can serve as an educational device; or they can tie in with a specific occasion (office anniversary); or special time of year (a pre-tax season counseling day).

It's common and helpful to have certain promotional items or 'giveaways' on hand at any promotional event, something useful that can serve prospective clients as a reminder of who you are, where you are, and what you do. Such 'giveaways' should have your logo, name, or other form of identity printed on it so that prospects can get in touch with you. Aside from business cards and brochures such items might include custom printed calendars, pocket calendars, bookmarks, ball point pens, key rings, rulers, plastic clipboards for notepads, refrigerator magnets, and so forth. But before you order such items, think of something that is useful, relates to your practice, and does not call for any great investment of money.

Here, then, is a sampling of some promotional endeavors worthy of your attention. Depending upon your talents, skills, and inclinations, you might consider trying one, some, or even all of the following measures at one time or another.

1. Announcing the opening of a new office, or change of address, or a new partnership, or expansion of facilities. Any of these can rightfully lead into a promotional event where you send out announcements or invitations, have an open house, provide refreshments, offer advice and information—and afterwards you can even write up the event as a press release.

2. There are all kinds of fairs and exhibits where you—as a tax professional—may participate: community fairs, neighborhood fairs, business fairs, career days at a college; or educational fairs, where the public comes to learn about various professions and professional services.

   If you become involved in such appearances, even occasionally, it may be worth your while to design an interesting booth or display to appear on portable panels that can easily be assembled and taken away. Use your imagination, but an innovative display can be a great conversation piece, and can attract a lot of attention from prospective clients. For example, you might devise a board that provides a brief historical overview of taxation in this country—with words and plenty of pictures. Or with the use of charts, diagrams, and photographs, you might show how the system of taxation operates. Or you could post various tax tips, regulations, and other informative and practical features. With time and practice, such presentations could easily evolve into a lecture you could give.

   In any event, whether you attend a fair or convention as an exhibitor, or even as a fair-goer, you are sure to make a number of worthwhile contacts there.

3. Teach a class or give a seminar. This could be done in your home or office; in a school or other community center. In this way you can inform the public about a variety of tax related matters: what a tax practitioner does, or what clients can do to help themselves; tax tips, financial planning, new regulations, and so forth.

   There may be certain public places—like libraries, colleges, or shopping centers—where you could situate yourself on certain days of the year, in order to answer questions and offer tax related advice—and thereby make the public more aware of your services and range of expertise.

4. Be a public speaker: contact libraries, community centers, religious organizations, and offer a public service lecture relating to tax tips, the effects of chang-

ing regulations on the taxpayer, and other aspects of financial planning.

If you are inclined to public speaking, then this is a skill you can gainfully employ in promoting your practice. Just make sure you prepare for your speech accordingly:

A. Prepare for your talk by familiarizing yourself well with all the pertinent material. Do the necessary research and be certain of your facts.

B. Thoroughly practice your speech and its delivery. Some do this into a tape recorder and then listen back in order to make corrections and adjustments.

C. When you deliver your talk, be aware of your posture, your voice and range, and be sure to make eye contact with your audience.

D. Keep your presentation on the short side, and be sure to leave over enough room for questions, discussions, and audience participation.

E. If someone asks you a question to which you do not know an answer—don't fake a response; simply answer that the question requires some research, take down the person's name and number and tell him you will contact him when you find out the answer.

If you lack confidence in public speaking but would like to give it a try, you might take a course in public speaking; then gain experience and confidence by experimenting in smaller settings: a high school classroom or a mini-seminar, and then slowly advance to larger gatherings. And then again, if public speaking is simply not for you, you can simply overlook this particular form of promotion and focus on the others.

5. In conjunction with business fairs and conventions, we mentioned the advantages of setting up a booth or panel display. Very often, such displays can stand on their own—open to public view—without your having to be there. Libraries, schools, recreation and community centers often have wall space or cases available for displays. If your display is interesting and informative you could get a lot of free exposure in such settings, just be sure to pin your own brochure to the display as well—and leave plenty of your business cards around.

Public Relations

Public relations is a difficult term to be specific about: on the one hand it does relate to your tax practice, but on the other hand, it goes far beyond the scope of your tax work. In a way public relations concerns 'the greater you', the 'community minded you'; or the 'you' that extends and interacts with different constructive facets of the community around you. It may involve coaching a little league team, or volunteering time to an Ambulance Corps. Such further extensions of self may or may not relate to or involve your tax work, but these activities somehow reflect upon your career as well—as if to say, that your professional and personal life are in some ways inseparable.

The suggestions that follow fall more easily into the category of 'public relations' than they do into the other headings listed above, so we discuss them at this point.

It is a long-standing and proven rule that one of the best ways for a tax professional to build his practice is to get involved with his community. Many professionals claim that there is no better way for a tax specialist to make a significant contribution to his community—then by way of his particular expertise; at the same time he establishes a network of friendships and professional contacts that will result in the development of a lasting practice. This sort of community interaction is usually classified as public relations.

1. Join organizations. For example, get involved with various non-profit organizations, the local Chamber of Commerce, the YMCA, religious groups or clubs, libraries, hospitals, and so forth. Of course, you don't have the time to join all of these, but try to connect with the groups you feel an affinity towards—"where your heart is in it" as well—and then make sure you can manage it in terms of your own professional obligations.

2. Become a sponsor or donor in various charity or fundraising groups. Many such organizations can use your help, your expertise, and will give you free publicity in exchange for your contribution, either in a printed program or journal, or in a newsletter, or publicly at meetings or celebrations. Such involvement on your part can also be the topic of a unique press release.

3. Participation in various community projects can put you in contact with individuals in other significant positions.

4. Try to attend various public functions and gatherings, the friends and acquaintances you make may have an impact on your practice.

5. Also, be sure to attend functions that relate to your own special interests: sports, hobbies, youth clubs—these, too, can be excellent channels for meeting new clients.

6. Welcome Wagons: Some neighborhoods, towns and communities provide a greeting service that distributes community information to newcomers in the area. Being a part of this service can also be a valuable resource for new clients.

In Conclusion:

As you can see from the previous pages, our concern here has been to present you with numerous ways of providing reliable information to potential clients in order that they should recognize the benefits of your tax practice.

In truth, however, we have only scratched the surface of a subject to which entire textbooks are devoted. Nevertheless, you do have sufficient material here with which to make productive headway in effectively marketing your practice.

Just remember: successful marketing—by presenting an honest picture of you and your service—can help you build a practice; but only you—by providing an overall quality service—can maintain the practice and keep it growing.

Chapter 5
PURCHASING CLIENTS OR A PRACTICE

Nothing is more inevitable than death and taxes.

Benjamin Franklin

Of life's two certainties, taxes are the only one for which you can get an automatic extension.

Jim Fisk and Robert Barron

A taxpayer—that's someone who works for the federal government, but doesn't have to take a civil service examination.

Ronald Reagan

Purchasing Clients or an Entire Practice

The majority of tax specialists are content to establish and build their practice by starting with a small nucleus of clients and gradually widening that circle through effective marketing and superior service. But there is a quicker, though more expensive way of establishing or expanding a tax practice—by the direct purchase of tax clients from other practitioners. This method of obtaining clients is employed by all professions, and generally works to everyone's satisfaction...assuming the buyer will provide service that is equal or superior to that given by the previous practitioner.

You may establish or expand your practice either by purchasing an entire practice, or just a group of clients from an established practitioner. Although individual clients may also transfer from one practitioner to another for various reasons (which we will soon discuss), you must nevertheless be willing to pay a fair purchase price to obtain a group of clients or a complete practice all at once.

The most obvious reasons for selling an entire practice are the death, illness, or retirement of the present practitioner. But occasionally, a tax specialist will give up a practice either because he has taken on a full time position, or is moving out of the area, or is going into another field of endeavor.

Though it is more expensive, this approach may have certain advantages over launching your own practice: you have a fixed clientele, an established location, and very often, a full inventory of supplies and equipment—perhaps even an experienced employee. And if the owner is anxious to sell, you may get a very good price as well. Buying an ongoing practice can also save you time, energy, and money you would otherwise spend establishing a practice. Nevertheless, you have to be wary of such encounters, and there is a fair amount of research you will necessarily have to do.

You need to know:

A. What condition is the practice in? A healthy state or one of deterioration? What has been the business trend for this practice over the last 5 years?

B. The practitioner selling may praise his own operation up to the skies in order to get the highest selling price, but what are his real reasons for selling the practice?

C. What about the general neighborhood; is the area growing or declining?

D. What kind of reputation or public image does the practice have—favorable or unfavorable? If it's positive, then you can capitalize upon it; but if it's a negative one, then you will have a lot of overhauling to do.

Also; be sure to get good advice from reliable friends or qualified professionals, i.e. lawyers and neighborhood bankers. This is not the sort of decision you want to make on your own; the cost and implications are simply too far-reaching. Then, after a thorough investigation, if major considerations check out favorably—well and good; but if not—then it would be strongly advisable for you to look elsewhere.

Should you Purchase an Entire Practice?

If you are relatively new in the tax field you ought to think carefully before taking on a full practice all at once. For one thing, it calls for a rather substantial investment of money; suggested purchase prices will soon be discussed. And then too, you may not at first be able to handle some of the more complicated returns; or your clients may sense that you are unsure of yourself, and coming from a seasoned practitioner, they may go on to look for someone more capable. On the other hand, an established, experienced practitioner incurs less risk when taking on a full practice...if he can handle the additional workload.

Our suggestion then—if you are just starting out—is to look for a practitioner willing to sell a group of clients, preferably small or middle income returns or for a part-timer with just a small number of clients. Even when a larger practice is up for sale, the practitioner may be willing to break his clientele into segments; transferring the larger clients to one buyer, and all or part of the smaller clients to someone else.

If you are interested in pursuing this method of establishing or increasing your practice, then you will need more information on how to...

1. Locate a prospective seller.
2. Establish contact with the seller.
3. Negotiate a mutually satisfactory purchase price.
4. Effectuate a smooth transfer of clients.

We will now discuss each of these steps in detail.

Finding a Seller of Individual Clients

Experience has shown the following to be likely, prospective sellers of individual clients or groups of clients:

1. Semi-retired tax practitioners.

   Very often, as a tax professional approaches retirement, he will gradually cut back on his practice by eliminating smaller, less lucrative clients. Yet these clients can be a very welcome addition to your developing practice. Bear in mind, however, that before seriously discussing the transfer of these clients, the practitioner will want to be sure that you are professionally capable of handling them, so it is important that you are able to assure the current practitioner of your professional skills.

2. CPAs cutting down on individual tax clients.

   Many CPAs or even large, non-CPA accounting firms become so busy—with management consulting work, electronic data processing, pension and profit sharing plans, and other specialties—that they simply lack time and personnel to handle individual, or even small business clients. While many of these firms are reluctant to accept any such new clients, they are somewhat committed to a number of these 'return' clients from previous years. They would be only too glad to transfer such clients to another practitioner—one who is able to provide quality service. Here too, before transferring clients, firms will want to be certain that they are moving into capable hands.

3. Attorneys.

   Although most lawyers are not expert in tax law or procedure, many, in the early years of their practice, do tax work to creatively occupy their time and bring in needed income. As their law practices develop, however, they generally reach a point where they prefer to divest themselves of all tax related work. If you happen to contact them at the right time with the right offer, you may be welcomed with wide open arms.

4. Real estate brokers and other non-professional tax preparers.

Almost every community has real estate or insurance agents, or other individuals who do some tax work on the side, but who have never really developed a sizable practice. Often, these individuals decide, or can be persuaded, to part with a portion or even all of their clients, so they can fully devote themselves to their primary occupation. Even those who are not quite ready to give up their practice find that some clients have simply outgrown their ability to accommodate them, and they would be glad to transfer these clients to someone more expert than they are.

Your next question might be: How do I find out if and when any of the above have clients for sale?

If you are in touch with other tax specialists, chances are you will hear through the professional 'grapevine' about practitioners planning to prune their clientele. And even if you are not in direct contact with tax preparers as a group, you may have an accountant friend or relative who you can confidentially ask to 'keep his ears open'. But there are other resources that can lead you to the acquisition of new clients as well.

1. Tax service and office supply men.

These salesmen are constantly in touch with tax professionals. They may even be among the first to know if someone is overburdened, ill, or otherwise wants to divest himself of clients. Let your salesmen friends know that you are in the market for clients, and they will be happy to make the contacts for you—especially if you let them know that there is some compensation in store for them.

2. Your own new clients.

A good indication that a certain practitioner or tax firm is giving up business may be the sudden appearance of new clients. You may at some point experience a flow of new clients, and all coming from the same firm. This should alert you to begin making discrete inquiries as to the reason behind the sudden switch: Is their current practitioner ill, or too busy? Is he changing his specialty, or moving into another field?

3. Finally, you may just try the direct approach of randomly contacting a number of tax firms, and other professionals, and respectfully inquire if they are in a position to transfer clients.

Approaching a Prospective Seller

Once you locate a likely resource for new clients, your next step is to approach him, directly or indirectly. Bear in mind that professional clients are not to be treated or traded like so many head of cattle, so your approach must be discreet and dignified.

One method is to send a friendly letter expressing your interest and asking the recipient to respond by mail, or phone, to set up a conference, if he wishes to discuss the matter further. The following is a sample introductory letter to this effect...

Dear ____________,

I am engaged in tax practice...at the following address, and I am currently interested in purchasing additional clients.

It has occurred to me that due to the growth of your practice, staff shortages, specialization on your part, or other developments, that you may wish to consider the sale of some of your individual clients.

If this is in fact so, please phone or write me so that we can set up an appointment at your convenience to discuss the matter further.

I can assure you that in serving your clients I would try to maintain the high standards of service that they have come to expect from you. Needless to say, your response and all negotiations would be held in strict confidence.

Sincerely,

John Doe

If you know the prospective seller personally (or if he knows you), and you are fairly certain that he is planning to drop some clients, then a phone call from you might be quick and more effective. However, be prepared to beat a hasty retreat if it turns out that you are mistaken. At that point you can politely close the conversation by letting him know that you would be interested in speaking further if he changes his mind, or if the situation changes in the future.

How to Locate Entire Practices Available for Purchase

As we mentioned at the outset, the most common reasons for selling a practice are death, illness, or retirement.

When a practitioner dies, you will probably find out about it soon enough. We suggest that after a decent lapse of time, you ascertain the name of the attorney handling the estate, and advise him that you are interested in purchasing all or part of the deceased's practice. It is best not to approach the widow or children, unless you have a mutual friend who can delicately broach the subject with them.

When you hear that a practitioner is about to retire, your best approach would again be through a mutual acquaintance, or other go-between, such as a lawyer or banker.

Another resource for finding a practice up for sale is the classified column in city newspapers, and the hefty Sunday edition is usually the best source for such information. Professional accounting and tax journals, and related periodicals, also carry classified announcements of entire practices for sale.

Negotiating the Purchase Price

To arrive at a satisfactory purchase price, the question uppermost in your mind should be: How much are these clients worth to me? Regardless of how valuable or lucrative any client is to the seller, he will be of little value to you if a) he won't stay with you, or b) if for one reason or another, you won't be able to service him properly.

But assuming you have a list of clients who, as a group, you can expect will stay with you, you nevertheless have to anticipate some degree of drop-off in the first year or two. For one thing, some clients may have had intentions of switching to another practitioner for some time, and were just waiting for the right opportunity to do so. Others may have developed a strong bond with their previous tax consultant, and are not willing to adjust to your style. For other clients you may be too young, too old, too new, too busy, or what-have-you! The point is that a certain amount of drop-off is inevitable; it should not be cause for concern, but it must be taken into account when establishing a purchase price.

The most common and accepted method for arriving at a purchase price is on the basis of gross volume. You will generally find that the going rate is between 100-150% (on rare occasions even higher), of annual gross fees collected from the particular clients. However, you should insist, if possible, on a contingency agreement whereby you pay the full purchase price only for those clients that remain with you at least two or three years. Clients who drop out earlier are then paid for on a sliding scale. For example, if the agreed upon price is 100% of one year's gross, we suggest that the purchase price be paid over a period of two years at the rate of 50% of gross billing per year. So for those clients who drop out after one year you will have paid only 50%, while the full price is paid for clients who stay with you two years or more.

If the agreed upon price is 150% of the gross, try to stretch payment of the purchase price out over three years on a similar basis. However, the seller may insist on a two year maximum anyway, on the theory that if a client leaves you after two years, it is your fault, not his. Indeed, it's difficult to argue with that position.

A variation of the above method is to set a price equal to—from 100-150% of one year's gross fees, less allowance for estimated drop-off. Then, if the total gross fee earned last year from clients transferred is $10,000, you might propose a purchase price of $10,000 to $15,000, less 20-25% drop-off allowance.

Even if you negotiate a flat purchase price, try to avoid a one year all cash deal. Payments spanning two years, or even over one season will give the seller considerable incentive to 'deliver the goods', i.e. do what he can to insure a smooth transition, and urge his clients to stay on with you.

Implementing a Smooth Transfer

Your final step is to make sure your new clients are transferred to you with maximum efficiency and a minimum of inconvenience to them. This, of course, will require full cooperation from the seller. First, he should write a letter to each client advising that since he is no longer able to be of service, he has made arrangements with you—another practitioner—to handle the account, and has therefore transferred all necessary files and records to you. Although some send this notification letter as soon as the deal is closed, we suggest that you wait until the beginning of the season; this will minimize the chances of a client's looking for someone else.

A short while later, contact the client by phone or letter expressing your desire to continue the professional relationship he has enjoyed with the previous practitioner and suggest an early appointment. Prior to the appointment, carefully look over the client's files and familiarize yourself with his particular circumstances and tax situation. It will also help the transition immensely if you can get the seller (if he is available) to give you a briefing—in writing—on those clients with special tax problems, personal concerns, or idiosyncrasies, and that require special attention. With this sort of pre-planning, you can avoid a lot of aggravation later. Moreover, many clients will appreciate the fact that you took the time and trouble to acquaint yourself with their particular situation before meeting with them.

It's natural, too, that a number of clients coming to you from another practitioner will be somewhat uneasy at first. This is especially true if the previous relationship was a close and long lasting one. They will wonder whether you 'have what it takes' to service them properly, whether they can really trust and confide in you. In short: they want to be sure you measure up.

It follows then, that you will have to exercise extra care with these clients. Make sure that all work is meticulous, double check your figures, and allow extra time to become acquainted with their situation and tax concerns. If possible at first, do not charge them more than the previous practitioner—unless there has been some material change in the client's circumstances, or type of return. But once he has stayed with you a year or two, you will of course want to treat him like any of your other clients.

At the same time, if the previous practitioner's fees were above your level, you have no moral obligation to charge the client less. However, if and when you raise your fees, we suggest that you only increase these clients' fees proportionately.

Individual Clients

When you purchase individual clients, pay only on the basis of regular recurring fees, such as tax return preparation, payroll tax returns, etc. Fees collected for representing clients at audits should not be taken into consideration because these tend to be non-recurring. However, when buying an entire practice, you will—by the law of averages—earn a certain amount yearly in audit fees, even though not from the same clients. So don't haggle too much over the purchase price on that account.

When you buy an entire practice, the seller may also want to dispose of his office equipment, fixtures, and perhaps even the office itself. Tax-wise, it would be to your advantage to allocate as much as possible of the purchase price to the physical equipment and furnishings, and to minimize the 'goodwill', or price paid for the acquisition of clients. The reason is that the former is depreciable for tax purposes, while the latter is not. So try to arrive at a fair compromise, but don't let this consideration ruin the transaction, if it appeals otherwise.

Finally, if office equipment does become part of the sale, be sure to check it out carefully before the purchase: it may be of inferior quality, in poor working condition, outmoded, or otherwise unsuitable to your needs.

A Few Additional Thoughts Before you Buy

Regardless of how eager you are to take on a full fledged practice, and in spite of how attractive the 'package deal' may seem—make sure that you approach the transaction objectively and cautiously.

- Thoroughly check out the condition of the neighborhood, as well as the office facility. Speak to bankers and managers of other business establishments in the area.
- Since the practice has a history, files and records should be available to you for closer scrutiny. Get professional help to go over the records with you to get a more accurate picture of the condition of the practice.
- Determining a fair price is usually a problem: the seller tries to get as much as he can and the buyer tries to get away with little as possible. A fair price—agree-

able to both parties—lies somewhere in between and the way to arrive at that fair price is through effective negotiation.

Negotiation tends to be a stressful encounter, but it need not be if you prepare yourself accordingly—both in attitude and technique.

1. Approach negotiation as if it were a mutual problem-solving venture, where you are seeking a solution that will be beneficial to all parties concerned.
2. Plan and prepare for your meetings with the seller by becoming as familiar as you can be with the issues at stake. This is one instance where ignorance is not bliss; rather—be as knowledgeable as you can be about the condition and the facts relating to the seller's practice, and be as clear as you can be about what you wish to accomplish.
3. Begin by discussing issues you are both agreeable to, and as you move along—seeking solutions to where you differ—keep summarizing whatever progress you make, as a form of further encouragement.
4. If possible, try to have a neutral intermediary present to serve as a sounding board to facilitate the discussion.
5. No matter what happens in the course of the meeting; try to end on a positive, friendly note: leave options open so that you can meet again, if necessary.

Chapter 6
EXPANDING YOUR PRACTICE

A professional is defined, not by the business he is in, but by the way he is in business.

Anonymous

What is a Client

- A client is the most important person ever in this office...
  in person or by mail.
- A client is not dependent on us...
  we are dependent on him.
- A client is not an interruption of our work...
  he is the purpose of it.
- We are not doing a favor by serving him...
  he is doing us a favor by giving us the opportunity to do so.
- A client is not someone to argue or match wits with.
- A client is a person who brings us his wants.
- It is our job to handle them profitably—to him and ourselves.

(From poster displayed at the headquarters of L.L. Bean in Freeport, Maine. In the poster the word 'customer' is found in the place of client).

Once you have overcome the first hurdle—of developing a nucleus of clients—chances are you will want to raise your sights and look for ways to expand your practice. Depending on your goal, you will want to attract sufficient clientele to provide you with either a comfortable second income or a viable, full-time tax practice. Of course, if your aspirations are more modest, and all you wish to handle is a small number of clients, then you may have reached that point already and will not be interested in further expansion at this time.

But if your sights are set on a larger practice, you will find that growth will come from three primary sources:

1. Recommendations from current clients.
2. Through the normal influx of additional clients, for example, by advertising.
3. And by upgrading your level of service to current clients.

Your Present Clients as a Source of New Business

As we mentioned before, your most effective and prolific source for new business is your current pool of clients. Personal recommendations from satisfied clients are vitally important for tax specialists striving to develop a quality practice.

Needless to say, you can't expect them to recommend you to others unless they are fully satisfied with your service, integrity, and technical know-how ...to the point that—not only will they come back to you each year—but they will recommend you to others as well. We will therefore discuss various ways to satisfy clients and impress them with your knowledge, competence, and professional skills.

Retaining Your Present Clients

In planning for the growth of your practice, direct your initial efforts at retaining the clients you now have. Some practitioners go to great lengths to acquire new clients, while putting little effort into keeping their client base. Not only is it easier, but wiser, to do all you can to keep current clients satisfied, rather then spend time chasing after new ones. And the reasons for this are simple...

1. The law of inertia teaches that it is easier to retain a current client than to acquire a new one.
2. An established client is more likely to promote your services, than one who flits about from one tax consultant to another.
3. Most new clients tend to be relatively small, lower-fee taxpayers, while those who have been with you for some time stand a good chance of developing into larger, more lucrative clients.

Moreover, a client who has been with you for a while and with whose affairs you are familiar, requires less time and effort (all things being equal) than a new client with whom you must first get acquainted. In other words, you will spend less time on his return and earn a higher fee per hour—than with a new client.

How to Keep Clients Satisfied

Fortunately, it is not hard to keep a client satisfied. If you can display technical competence along with the right interpersonal skills, then you have a combination of talents that would be hard to beat in any field of endeavor. What follows is a summary of some of the more important features to keep in mind in your dealings with tax clients. A few of these points are discussed in greater detail in other sections of the text.

1. Always be friendly. Show that you appreciate the patronage and confidence of your client; whereas being arrogant or haughty is certainly not a good business trait. In fact, there may be no quicker way to lose a client than to exhibit this fatal characteristic.

2. Express an interest in your clients and listen attentively to what they have to say. The focus of any meeting should be the client and not you.

3. Be patient ...don't hurry. Impress each client with the fact that regardless of the size of his or her income, and no matter how busy you are, this case will be receiving the utmost personal attention, and that you will try to find all possible legal deductions and tax savings to benefit the client.

4. Be reasonable in your fees, and in your initial interview spell out clearly what your charges are and how they are calculated.

5. If possible, surprise your client by doing something extra or unexpected. For example, your examination of his financial transactions may bring to light certain deductions which were lost because the payment wasn't handled in the proper way. If you point this out to the client and explain how he could handle such matters in the future to get the greatest tax benefit, you will have taken an important step towards winning his confidence and gratitude.

6. Don't be a 'know it all'. Do not hesitate to check your reference material or indicate the need for further research when you are in doubt. Most clients realize that tax law is too complex to be stored in one person's mind. They will appreciate the fact that you are taking the time and trouble—in spite of your busy schedule—to make sure their return is being prepared correctly.

7. Impress your clients with the importance of keeping adequate financial records. The taxpayer who is in the habit of writing down all of his expenditures can be reasonably certain that no important deductions will be overlooked when the tax season comes around.

8. Make sure your clients retain all checks, bills, and other proof of deductible expenditures. Stress the importance of having all this information readily available in case of an audit on the return.

9. Always search for new ways to serve your clients. Some services will result in

additional income for you, while others will serve to keep your clients pleased—and will pay off financially in the long run.

10. If your client is fairly intelligent and knowledgeable in tax matters, make it a point to explain what you are doing and why. If it entails extra work or some risk of an IRS audit or disallowance, make sure to spell it out clearly to him. On the other hand, if the client does not know a thing about taxes—and cares even less—then there isn't any point in being overly explicit, except to point out the extra work or risks involved.

11. The strictest confidentiality (as we will soon discuss) is of paramount importance. However; there is nothing wrong in passing on information of a general nature (without any form of identification) of something you learned while preparing another return, and that may be helpful to this particular client. For example, you may have learned from other taxpayers that IRS auditors are zeroing in on particular deductions or income items; or you may have picked up some business ideas that would be helpful to your client. Passing on such tidbits can only enhance your reputation and the value of your client relationships.

12. When setting up an appointment with a client, especially a new one, regardless of how much free time you have—it's generally a good idea to suggest a convenient hour a day or two ahead, and not immediately, unless the matter is urgent. It's a well proven axiom that 'success breeds success'. You want to appear 'preoccupied' rather than 'unoccupied'. Nevertheless, don't give a client or prospective client a hard time getting an appointment, or he may take his business elsewhere. And pay special heed if the client describes the matter as urgent. Remember; the problem may seem basic to you, but it may be giving your client restless nights.

13. Try to impress upon your clients that they should consult with you before venturing into any sizable acquisition, business deal, or other financial arrangement; if only to consider the tax aspects and to minimize any unfavorable tax results. Too often, the tax advisor finds out 'after the fact' when it's too late to be of any help. You should actually try to educate your clients to think 'taxwise' all year round, and not only during the tax season.

14. To keep clients genuinely happy requires some insight—on your part—into their personalities. For example, some clients have a primary concern of keeping out of trouble with the IRS. They would much rather pay a larger tax bill and forego some deductions, so that the return will not be questioned. Others are ever ready to don their armor to do battle with the IRS, if they have some chance of winning. So when you come across 'gray area' deductions, tax saving maneuvers, or other techniques—guide yourself accordingly. There is no point in saving a client $100 if it will give him sleepless nights throughout the year.

15. Be tactful in your dealings with clients. Never joke about a client's income, his means of income, or about his losses. All of these are serious matters to him, and they should be to you as well.

16. Finally, pay particular attention and be especially sensitive to new clients who have never before employed a tax professional. They may be hesitant or embarrassed about disclosing their financial affairs, so be pleasant and understanding as you guide new clients through the tax procedure.

Personal Relations Along With Quality Service

Until now we have discussed practical ways of projecting your image as a competent, skillful, and dedicated professional. But clients are not machines, they are human beings who want a practitioner who is not only knowledgeable in his field, but also exhibits a sincere personal interest in his clients and does not relate to them as mere cases or numbers. In fact, when clients make overall judgments about a practice they take a wide variety of features into consideration—the appearance of the office and staff, as well as how they are treated—and it isn't just a matter of a practitioner's technical skills.

There is also another good reason to establish a healthy personal relationship with your clients. You will find that your stiffest competition is not from CPA's: their fees are too steep and they are usually too busy to aggressively seek new clients. The same applies—to a lesser degree—to non-CPA public accountants. Nor do you have much to fear from the small, independent, part-time tax preparer. As we mention elsewhere, many of these are inept, poorly trained, and hopelessly outdated in their work. Your technical training and knowledge gives you a decided edge over them.

The heaviest competition that most practitioners encounter is from the national tax services, who are always hungry for more business. Their main weakness, however, is their impersonal, hasty, assembly-line approach to tax service. This fatal flaw, perhaps more than any other, prevents them from taking over an even greater share of the tax market. And no doubt—this is primarily what keeps them from forcing out the capable, independent specialist.

Therefore—friendly, personal service and attention to client detail is perhaps the most powerful weapon you have in your battle with the super-giants. Use it well—to your and to your client's best advantage!

Use every possible opportunity to add a personal touch to your dealings with clients. Fortunately, there are many ways to show your personal concern, and if you make it a habit to seek out such opportunities, you will build a strong personal relationship with many clients, and this will lead to the development of a successful and gratifying practice.

Here are just a few ways whereby you can add that personal touch in your dealings with clients. It is by no means an exhaustive list, and you can certainly add to it with innovations of your own.

1. Be a courteous host in your office: Greet clients at the door, and hold the door open for them. Offer to hang up a client's coat. Seat your client down before you seat yourself. And at the end of the appointment, escort your client to the door.

2. Take note of pertinent, personal information regarding a client: names of family members, a vacation he plans to take, a class he attends, or a hobby he indulges in, etc.

   Keep a written client file to help you record such information, and to use for reference prior to any appointment or call you make. That way, you can ask about his family, the vacation taken, or other events that transpired.

3. Make sure you refer to and pronounce your client's name correctly.

4. In your client file keep a record of special dates and events, and on such occasions—birthdays, anniversaries, graduations, weddings, births, etc.—be sure to send a personal card or note.

   - Send seasonal greeting cards.
   - Send letters of appreciation after a service is rendered.
   - When you hear that a client or some member of his family is not well, be sure to send a get well card along with a personal note. And if at all possible, try to pay an in-person visit.
   - If you come across a news item or other information of special interest to a client, pass it along to him.
   - And at any other momentous occasion—buying a new home, opening a new business, or having won an award—be sure to send along a congratulatory message.

5. Have light beverages or snacks available for clients, and offer it to them. For children, have candy on hand or other little knickknacks to give away—to occupy their attention while you meet with the parents.

   If you remember a client's preference in a beverage or refreshment, try to have it available for when you meet.

6. Be informative as you discuss professional matters with clients. Take the mystery out of what you do by providing clients with a better understanding of certain terms and procedures. This will make them feel more at ease, more comfortable with you, and more confident with you as their practitioner.

7. Make your client feel special and important however and whenever you can.

   - Compliment clients sincerely. Develop a genuine interest in what they do.
   - An appointment with you should be a boost in their day, a lift to a client's spirits and self esteem. Clients should leave their meeting with you feeling optimistic and good.
   - Let clients know that you are thinking about them. Make out of season calls just to see how a client is doing.
   - Also—encourage clients to keep in touch—with everything from questions to criticisms.
   - Try to respond to a client's calls or written correspondence promptly.

8. If it's a new client, do something extra, for example, ask if he or she needs directions to your office and explain precisely where parking is available.

9. If you cannot be present for an appointment, excuse yourself beforehand and offer some sort of explanation. But don't just fail to show up with no prior explanation.

10. Many clients may be neighbors of yours, or fellow members in a community club or organization, people you have occasion to meet throughout the year. Undoubtedly, opportunities will arise to render small favors to these individuals—especially if you keep your eyes and ears open...a lift in your car, the loan of a book or tool, or an offer to help with some small chore. Such gestures can easily turn an acquaintance into a loyal friend, and even into a promoter of your professional services.

11. Office hours: Being available at times when clients need you most is an excellent way to satisfy clients and generate new business. It could be that many of your clients (and prospective clients) are occupied during conventional business hours, and would find it both helpful and convenient if you could meet with them evenings, week-ends, (even mornings)—or perhaps even in their own homes or offices.

    If you can accommodate them—in terms of time and place—you may find this to be a key selling point in developing your practice.

Follow-up System

Another important point to keep in mind when preparing a return is to immediately make a list of items to consider for next year's return. This includes carryovers (such as capital losses), contributions, investment credit, special notations, and memos regarding items to be aware of: i.e., loose items such as the sale and expected purchase of a replacement residence, copies of estimated tax paid, and items affecting the basis of property to be depreciated. Also, in this file, make note of any special problems you had with the client, or special items, or peculiar deductions that were questioned by IRS in previous years and that are likely to recur. Another item often overlooked: when a client or his wife turn 65, he or she will be entitled to an extra exemption. Put a "tickler" into your future file to insure that this won't be overlooked next year.

An excellent procedure—one that will repay your efforts may times over—is to institute a follow-up file or calendar for items to take care of during the year. One obvious example would be to remind the client to make his estimated tax payments on time. In this connection, it pays to review his record before the end of the year to see if sufficient estimated tax payments have been made. If there is a sizable underpayment, and the client is employed, you can save him some penalties by having his employer withhold additional tax before the end of the year. Any tax withheld is treated as if it had been paid equally throughout the year and will reduce or eliminate any underpayment penalty.

Another personal—and effective—expression of professional concern can take place when you become aware of legislation pending that may affect your client. If this legislation is enacted, why not get in touch with your client immediately to discuss what can be done on his behalf? He will appreciate—and let his friends know too—that you don't get this type of individualized treatment from the large, national tax services, not to mention the IRS.

Another way to show personal interest is to forward an occasional news item or magazine article relating to a client's particular business, side interest, or hobby.

If you do employ assistants, it's a good idea to assign each client to an individual who can then become familiar with that client's problems and concerns. Of course, you run the risk of building too close a relationship between the two. In the event the assistant leaves you and sets up his own business, he may take the client with him. On the other hand, we feel the benefits outweigh the possible dangers. One way to minimize the risk is to make sure that you also meet with the client occasionally; at least give him a call or otherwise maintain the relationship—letting him know that you are there and well aware of what's going on with him.

The Confidential Nature of Tax Work

In the course of your tax work, clients will confide in you business and personal matters that they wouldn't tell their wives, children or even their best friends. You will probably never cease to marvel at the varied and unexpected bits of information that are revealed to you.

Needless to say, this special, privileged status imposes the burden on you of keeping all such business and personal matters in your strict and inviolate confidence. Never—but never—discuss clients' affairs with anyone. The slightest hint or suspicion that you are a gossip can do irreparable harm to your entire practice. Likewise, a well earned reputation as a tight-lipped professional who does not—advertently or inadvertently—divulge any kind of information, can be very advantageous to your practice.

Every tax practitioner has at one time or another been subject to adroit questioning by individuals trying to elicit personal or financial information about relatives, friends, neighbors or competitors. When this happens to you—as it surely will—your best defense is to be completely uncooperative. Your interrogator will soon get the message, abandon his fruitless efforts, and acquire new respect for you.

Never assume that a client's wife, relative or business associate is acquainted with his business affairs. For instance, you may be preparing Mr. Y's tax return and find that you need to know whether some stocks on which he reported dividends are owned by him alone or belong jointly to him and his wife. You try to call but he's not at home, and the filing deadline is approaching. Don't ask the information of Mrs. Y. It may be that she is not even supposed to know that her husband owns stocks.

In this regard, it is important that you acquire the habit of never leaving any completed or incompleted tax returns, or other working papers on your desk where others can see them. Whenever a return is finished or temporarily set aside, always put it into a closed file folder. Many tax specialists make it a practice to not even divulge who their clients are. You may also come across individuals who, for one reason or another, do not want it to be known that they are coming to you.

Confidentiality is extra important if you serve several competitors or are trying to convince your clients to refer their competitors.

Creating a Public Image

A most valuable asset to develop, and one that will engender client loyalty, is to present yourself as a meticulous and responsible executive: you keep your word, you meet appointments on time, and you fulfill your promises.

Here are a few practical suggestions to help you create and reinforce such a 'public image' of yourself.

- Immediately confirm every appointment in writing. For example: Tuesday, a new client phones you and makes an appointment for next Monday at 4:30 P.M. As soon as you get off the phone, send this client a card confirming the time, the place, and advising him of what to bring along. Aside from the actual reminder value (which in itself is worth the effort), you have immediately established yourself as a person who is meticulous in every detail.

. You promise to mail a form or paper to a client by tomorrow morning. Try your best, even if it's greatly inconvenient, to keep your word and have the item in the mail as promised (or don't make the promise). More important than the document or form, is the fact that you are dependable. You must become known and accepted as reliable, both in tax knowledge and in fulfillment of your promises.

- Make sure to be on time, or even a few minutes early for every appointment. This shows the client that you consider the meeting important, not a secondary or incidental affair.
- Carry your briefcase and handle your working papers with obvious care. Never allow the client to get the feeling that it's 'just another piece of paper' you're handling. Remember that your service consists of far more than just filling out forms. . Even if tax work is only a part-time occupation, don't allow your work or attitude to reflect a feeling of unimportance. Remember: if you attach importance and dignity to your work, so will your clients. On the other hand, if you treat your work casually, your clients will assume the very same attitude.
- Never pass along to a client a paper or form that contains any error, even of a trivial sort. Your office must convey your accuracy and attention to even minor details which have no direct bearing on income tax work. Watch your spelling and grammar when preparing letters, reports or statements.
- Bear in mind that the average client has no way of assessing the technical capabilities of his Tax Consultant. He often judges the total scope of his consultant's knowledge and expertise by observing how he handles other details: Do his papers look professional or sloppy? Is her well organized or is he hopelessly off-schedule; does he have confidence in what he does, or not? Therefore, these 'external' concerns carry an added measure of significance.
- While tax laws are too complex to demand that you have all the answers at your fingertips; it is important that you be sufficiently well versed, so you can discuss common tax problems with confidence. This is not to say that you cannot inform a client that you are in doubt over a particular point or that a matter requires further research. On the contrary, such honesty can build confidence between you and your client. But generally, you should be familiar with your subject and be prepared to answer basic questions regarding his return.
- You will be doing business with many personal friends, neighbors, and acquaintances, and it becomes all-too-easy to meet with such clients in informal settings, or in comfortable, informal attire. Nevertheless, if you desire to be perceived as an experienced professional, it is advisable to dress for meetings in

business attire This is especially important when meeting with new or prospective clients. Also in this connection, we should mention again, that the place you select in your home as a conference area, should not interfere with other household activities. You can well imagine that a run of personal phone calls, a parade of people, or a blaring TV will do little to create an air of professionalism in your client's mind. However, a room arranged with business in mind—set apart from others, and with explicit instructions to household members that you are not to be disturbed during business hours...will impress your client and set the tone for the importance of the meeting about to take place.

A wise person once remarked; "Almost everyone looks at you with the same eyes with which you look upon yourself". If you consider your work important, if you dress accordingly, speak accordingly, and act accordingly—it is almost inevitable that your clients will view you, and your work, in the same light.

How to Handle a Client's Tax Questions

You will find that clients will oftentimes visit or call to ask for your advice on a pending or completed transaction or business deal. It may not seem important at the time, but the manner in which you handle the inquiry can have a lasting impact on your relationship with a client, and this, in turn, can affect the growth and development of your practice.

Here are some suggestions from experienced practitioners that can help you improve the quality of such client exchanges.

1. Make sure you have all the facts: The client may not be aware of all the facts he has to consider. He may ignore a certain detail thinking that it's trivial; yet it may be vital to arriving at a proper assessment of the transaction. Therefore, you have to make sure you gather all the necessary information. Remember, if your advice or opinion proves wrong, your client may put the blame on you—not on himself.

2. Give clear-cut answers, don't hedge. Take all the time you need to consider a problem, but once you have an answer—state it clearly and concisely. And if you are in doubt, say so. Even the most experienced professional is not embarrassed to admit that he does not know the answer to a particular question.

   If the concern is one of a tax rule interpretation, make sure you inform the client of all possible, or likely interpretations.

3. Consider the client's tax knowledge in your reply, don't try to impress a client who is only vaguely familiar with tax law and procedure with a long detailed account of how you came to your conclusions. In other words, gear the amount of detail you provide in your dealings with a client to his particular level of understanding.

4. Try to offer alternative suggestions. If you must advise or caution against a proposed financial move, at least try to devise a way of accomplishing the same objective—with even more advantageous tax consequences. Nothing will endear you to a client more than your ability to—legally—find your way around a specific obstacle.

5. Finally, prepare a written memo of every important discussion, and send a copy to your client.

   Recording your recommendations will serve as a safeguard against misinterpretation, will help crystallize your thinking, and—above all—will provide a clear record of all that has transpired.

   In writing, briefly summarize all the facts that bear upon the situation, set out the details of the problem, and then state your answer or any alternatives you suggest. Incidentally, a full written account will certainly help you to justify a fee for your consultations, or a larger fee at tax return time.

Think of quality service as a product:

1. It's easy to think of a purchased product as something that either works well, tastes good, or is attractive in appearance; but you can also picture a 'service' as a type of product.

2. Service is produced and delivered as soon as—and for as long as—you are in touch with your client.

3. Service is 'experienced' by the client for as long as he/she is in touch with the practitioner or his office help—either in person, over the phone, or through the mail.

4. Like an electrical circuit—when there is contact between practitioner and client there is a flow of service that consists of solving tax concerns and satisfying personal expectations.

   In a survey designed to find out what clients expect from 'financial service' professionals, researchers discovered that clients are concerned with the following...

   A. Reliable help: in areas where most clients feel they do not understand the complexities and subtleties of certain financial procedures, they want to work with someone who is competent and trustworthy.

   B. Equal treatment: clients—regardless of how big or small their income—believe that everyone should be treated equitably, in terms of time and attention.

   C. Response to individual needs: clients expect prompt and continued assistance, especially when problems arise.

D. Reasonable rates: clients expect that fair fees should yield good service.

Personal Appearance

Considerations of personal appearance are crucial to success in any profession, but are especially important when you provide others with a service; because in a professional service—you, so to speak—are the 'product' that customers look at and assess.

Personal manner and appearance convey to others important information about your own sense of being. Try to become aware of how others perceive you; and then—if necessary—take whatever reasonable steps you can to develop an image that enhances your professionalism. The rule is: if you don't think of yourself as a professional, and act like one; then others won't either.

Here, then, are a number of points that relate to manner and appearance.

Be mindful of your appearance both in and out of the office. Make sure your clothes are neat, clean, and pressed; and your shoes shined.

In terms of style, a conservative selection of business clothes is a safe choice to ensure a professional image, as opposed to unconventional or loud colors and styles. Remember, accountants and tax practitioners, like bankers and lawyers, are expected to be on the conservative side.

Be aware of how you walk, sit, and carry yourself. Good posture is important, it communicates confidence—so sit up, don't slouch.

Eye contact and gestures are important. You should have normal, casual eye contact, but don't stare at clients; it will make them uncomfortable.

Smile when appropriate; a pleasant welcome smile is a good ice-breaker, it also transmits warmth and confidence.

Your voice, too, should elicit professionalism, be relaxed and natural, but speak clearly and distinctly.

Office Appearance

Your office or work space should be arranged in such a way that it enables you to get your work done; it should be pleasant enough so that you can spend many hours there at a time; and it should be neat in appearance and comfortable enough, so that visitors feel welcome and at ease in it.

If you have a waiting area, you should supply it with comfortable furniture and reading material—or other diversions—so that those who are there can well occupy their time while they wait.

Considerations of manner and appearance extend to all facets and phases of your client's contact with you. Therefore, you must see to it that it is a favorable experience every step of the way. This extends to first telephone contact...the sight of your office...the manner and appearance of your staff...office interactions...billing procedures—all of these should leave a positive, professional impression. There is a business adage that runs—'you don't get a second chance to make a good first impression'. So in addition to being a competent tax practitioner, you must be able to oversee your professional image and service as well. But it's challenging...it's stimulating...eminently gratifying and deeply rewarding.

In Conclusion

Since this chapter contains such essential information, it pays to emphasize and reiterate certain points in order to raise your level of awareness regarding the importance of quality service, so here are a number of closing considerations worthy of your attention and periodic review.

1. Be honest. Never try to fool a client or oversell your capabilities.

2. Exhibit integrity: be dependable, keep appointments, and keep to your word. Otherwise, don't make promises that you cannot fulfill.

3. Take pride in your work. Approach your practice with an air of respectability and responsibility. Treat your time and efforts seriously—as if your business and reputation depend upon it—as indeed they do.

4. Be patient with clients., spend extra moments with them if and when you can—answering questions, informing and explaining. These extra moments are an investment that will pay off in clients that will remain loyal for a lifetime. In this respect, remember, quality service does not cost—it pays!

5. Exhibit both depth of knowledge and breadth of knowledge.

   Know your own profession well. Be ever on the alert for changes in tax legislation or other information that may affect your services to your clients. The more you know about your own subject—your own tax business—the more professional you are. This is depth of knowledge.

   You should also exhibit a sincere interest and understanding of the work involvements and activities of your clients. Through reading and a little research, become familiar with the basics of what they do. Not only will this enhance your own understanding of the world, and improve the quality of your client relationship, but it will also enable you to service him better professionally as well. This is what is meant by breadth of knowledge.

6. Train your staff well. They should be competent, courteous, and a reflection of all the policies that you are particular about. Make sure they are informed

about the services you render, well trained regarding their own tasks, and familiar with general office procedure, so that work and communications can flow through your office effectively and efficiently.

7. There is a basic difference between a business which is product oriented and one—like a tax practice—which is service oriented.

A product—whether it's a food, furniture, or a house—is something tangible, you can smell it, see it, and put your hand on it. But a service, very often, regardless of how great the client benefits may be, falls more readily into the realm of intangible, the immeasurable, and it's harder for clients to perceive the time and effort which are expended on their behalf.

Therefore, the quality of your services—your appearance, your office appearance, the courtesies and amenities, the way you treat and remember clients—all these features play a more pronounced role in many service professions, because these are aspects of your practice that are more 'tangible'. These are aspects of your practice that clients can more easily relate to, understand, measure, and even put their finger on.

These considerations should, of course, in no way infringe upon the competency and skill of your technical know-how. It is only meant to point out the importance of quality service. Perhaps we could say that handling the intricacies of taxes well is more the 'mind' of your practice. Both are necessary if you are going to project an overall positive image of a tax professional.

8. Remember, when you focus on current clients, and gain their trust, confidence, and satisfaction, they will not only remain loyal to you, but will also recommend you wholeheartedly to others.

Chapter 7
TAX RETURN PROCEDURE

The first step in preparing a client's income tax return involves the client interview. During the interview you should try to gather all the data you need to prepare the return, and thereby minimize any time you might spend backtracking or requesting additional follow-up material. For this reason many practitioners prepare a special form listing all the items they might need, which they have on hand at every initial interview.

Make it a practice, then, to always process and complete all tax returns as quickly as possible—without sacrificing the quality on which you build your reputation.

If possible, have your clients make an appointment before coming. This will save them waiting time and it will put you at ease as well. Get a desk calendar or an appointment book, and mark down the day and hour of any scheduled meeting. Obviously for those cases which involve more extensive transactions, you will have to allot more time.

Many consultants do not wait for their clients to contact them every season, instead, they write or call their clients in advance, and set a convenient time to meet. Bear in mind, though, that except for farming areas you will not be able to make appointments if it is too early in the season. In fact, if your reminder letters go out too soon you increase the chances of your clients not responding to you, so you have to learn how to precisely time this sort of correspondence.

Client Information You Will Need

When making an appointment with a client, advise him in advance of the records, documents, and figures you will need to properly prepare the return. Here is a general list of most (but by no means all) of the documents you will need:

- All W-2 forms received from the taxpayer's employers.

- All 1099's, 1098's etc.
- If the client owns rental property a list of all rental receipts and expenditures.
- Complete itemized list of personal expenditures such as medical expenses, contributions, taxes and interest paid, casualty losses, child and dependent care expenses, alimony payments, and job related expenses (educational, work clothes, union dues, travel, transportation, and moving expenses, etc.).

In cases where a client sold any stocks, bonds, real estate or other property, you will need complete information such as broker's slips, or closing statements, etc. Also request any and all information regarding the original cost and date of purchase. You must impress upon the client the fact that the more complete his figures and records are, the better prepared your return will be, and the more money you may be able to save him in taxes.

Also instruct your client to jot down doubtful items, and to discuss them with you. Tell him to include any item that has even the remotest possibility of being an allowable deduction. Such a list may point to sizable deductions that are often overlooked because the taxpayer assumes they are not deductible.

Bear in mind that many clients are apt to forget or overlook deductible expenses, so you should really regard this list as your starting point for a more exhaustive search of such items. A thorough checklist (as found in most of the popular tax guides) will be very helpful for this purpose. If you take the time and trouble to make a careful inquiry for all legal deductions and tax saving possibilities, your efforts will be well rewarded

In the case of a professional person or of someone in business, you, of course, must have complete records or statements of all business income and expenses. If you can get such a list from your client, it will simplify matters considerably. Many taxpayers, however, keep scant records, if any, and you may have to reconstruct their deductible expenses from check stubs, bill, and vouchers. Naturally, you have every right to charge for the extra time and effort involved.

If your client has the necessary records, it is not necessary that he bring along canceled checks and receipts to verify the various deductions. However, for reasons we will soon discuss, we recommend that you ask him to indicate next to each listed deduction whether it is based on a check, receipt, other form of proof, or simply on his own estimate. If the item is an estimate, ask for some brief notation of how he arrived as the stated figure.

Finally, you should ask every new client to bring along a copy of last year's return, and if available, of previous returns as well. This is especially necessary in the case of a taxpayer taking a depreciation deduction, since the depreciation rate should be consistent from year to year. Earlier returns should also give you a clue to additional deductions the taxpayer may be entitled to, or they may help you to point out possible deductions or tax saving opportunities that were missed. In the case of previous clients, you should have your own copies of previous returns to refer to.

The importance of this habit cannot be overemphasized, and it will give you a valuable advantage over more experienced and better known competitors. Many accountants and tax preparers just ask the client a few questions and fill out the return, without bothering to make a thorough search for tax savings. They thereby fail to give the client the full benefit of their expertise, and they also miss out on a possible fee, and practice-building opportunity as well.

Should you make use of a questionnaire? Many tax specialists prefer to use a pre-printed questionnaire when interviewing clients to reduce the likelihood of omitting important items. Others feel that such a questionnaire gives the meeting an assembly-line look, and detracts from the all important personal touch.

Some practitioners have found a compromise, they distribute questionnaires to clients before they come in. This helps them gather the necessary information, assuring them that nothing has been overlooked, while it facilitates the return procedure. Mailing such a questionnaire to all previous clients at the beginning of the season, along with a friendly cover letter, can be a helpful reminder and also serve to invite the client back—and hopefully, early in the season.

Always use tact in questioning a client. Instead of asking, "Are you still married, Mr. Jones?" you could say, "Your wife's name is Betty, isn't it?" Or rather than ask whether a client's sick mother is still alive, say something like, "Now last year you claimed an exemption for your mother..."

Another instance that requires delicate handling is the case of female clients nearing the age at which they are entitled to an additional standard deduction, or they may be affected by one of the other age-related provisions. Instead of asking, "Are you 65 yet, Mrs. Brown?" simply ask for her birth date and do the figuring yourself.

In other words, in certain sensitive areas, it pays to be discreet.

It is generally a good idea to quickly run through last year's return. You then become familiar with the client's tax picture and you reduce the risk of omissions and oversights. It also helps to point out and explain to the client any necessary changes due to new or changed regulations. Suppose an important deduction always claimed by your client is suddenly eliminated due to legal changes. It is better to show a client who has always claimed this deduction that he can no longer do so, than to explain this fact **after** he questions the unexpected jump in his tax bill.

Often, during the preparation of a return or even before you start, a client may ask: "Well, how does it look? Will I have to pay additional tax or will I get a refund?" And unless the situation is quite obvious, do not hazard a guess. You may have to eat your words. This is particularly appalling when you assure the client that everything is fine and he can count on a nice refund, then you find yourself having to inform him that you miscalculated, and instead of a refund he has to fork over additional tax. Do all you can to avoid such embarrassments.

Preparing the Return

Once you're sure that you have all the information you need you can start on the return. Indeed, if you use a computer you could begin to enter the information even before and fill in as you go along. But you'll generally find it more efficient and less error-prone to assemble all the data first.

On the other hand, if you prepare returns manually (and a surprising number are still being done that way) you have little choice but to hold off until you're all ready. You don't want to risk having to do an entire return over because of one missing or overlooked piece of information.

No matter which method you use, look over last year's return (or the 'pro-forma', if your software provides one) to make sure that the basic client information has not changed, or to note any changes. Then carefully go over the client-provided information.

For manually prepared returns we recommend the following procedure.

Fill out the first draft of the return using the form and method you have chosen. Go over all calculations and check carefully for errors. Such mistakes, if carried over to the final return are particularly embarrassing. If you are filling out the form by hand, using a pencil for this copy will facilitate corrections and erasures.

Make an exact copy of the corrected first form. This is the copy you file with the IRS and it should be neatly written in ink or typed. Attach all necessary schedules, statements, and W-2 Forms, if your client is employed.

As you gain experience, you will find it more convenient to prepare the simpler returns directly on the tax forms. For the more complicated ones, you may first want to assemble the data on a worksheet and then transcribe it directly to the return forms. However, we suggest that in preparing more complex returns, you prepare a pencil draft first, to facilitate changes and corrections.

Remember: When your return is neat and legible, it gives it a professional look, and it also promotes accuracy, and helps reduce IRS audits and questioning.

Most tax practitioners prepare three copies of each return, accompanying schedules, and the various attachments. One set is for your own files, one is for the IRS, and one is for the client. You having a copy is a safeguard in case the client loses his. Such a procedure is therefore highly recommended.

There is another important advantage to having a copy of every return available in your files. Very often, a client will call you with an urgent question that he needs answered immediately. He may have received a routine IRS inquiry, or have a simple tax problem, or he may want to know why all of his fellow employees received refunds while he paid an additional tax (he already forgot about the $1,500 extra income he had last year, from which no tax was withheld). You will agree that a response of..."Hold on a minute while I get your file" is preferable to ..."Well, I don't know; I would first have to see your return."

With copy equipment being so efficient and effective, most practitioners opt for this method over the use of carbons. However, there is nothing wrong with using carbons, just be sure to use a high quality carbon paper and insert it correctly. Incidentally, any good quality copy is acceptable for filing, as long as the signatures are affixed after the copy is made.

Before filing the return or delivering it to the client, make sure you proofread it carefully. If you have more than one person preparing returns, it's a good idea to have the proofreader be someone who did not prepare the return. This will automatically serve as a check on the preparer's work.

A note of warning: Some practitioners feel that computer-prepared returns do not have to be checked over. Unfortunately, this is not so. Computers are unforgiving, and the slightest misstep or omission can easily result in a return that's off by hundreds or thousands of dollars. So exercise caution—especially until you've developed the 'feel' that alerts the experienced tax professional when something is amiss.

The proofreader should also be instructed to make sure that all required schedules and attachments (such as wage statements, supplementary explanations, etc.) are attached to the return. You do not want to have pieces of the return floating around your office after it has been filed or given to a client.

One of the more common errors that occur in filing a return is the inadvertent omission of one or more of the supporting schedules or statements. This triggers a delay in processing the return and it may, occasionally, even trigger an audit. One way to avoid this problem is to inconspicuously number all forms and attachments in sequence. Then, when collating the return, it will be easy to see if one or more of the numbered sheets is missing.

While most practitioners mail the completed return to the client for his signature, some personally deliver the return, while others ask the client to pick it up. The last two methods enable the practitioner to collect his fee when handing over the return. On the other hand, doing so may mean spending valuable tax-season time in conversation with the client. Nevertheless, where a client is habitually late in paying his bill, this procedure should be considered.

When delivering the return to the client, make sure to give him explicit instructions as to what he should do. Tell him where to sign, where to mail it, and when. Tell him whether a check is to be attached, and if so, for how much, and how the check is to be made out. If it is a joint return, make sure the wife signs it too. Many practitioners use a written or printed sheet containing these instructions. If the return is mailed to the client, rather than handed to him, such an instruction sheet becomes more of a necessity. There is a sample of such a sheet in the Appendix of this text.

Most preparers ask each client to mail his own return to the appropriate tax authority. This puts the responsibility squarely where it belongs—on the client. On the other hand, if you undertake to mail the return, and for some reason it does not arrive; or the client does not receive his refund fast enough, he may suspect that you delayed, or never mailed the return.

However, we do suggest that you follow up a few days before the filing deadline—by phone or by mail—to make sure the return actually went out. Clients will also appreciate the personal concern on your part.

A practice followed by some practitioners and one we highly recommend, is to not only provide pre-addressed envelopes (typed, rubber stamped, or printed) for mailing the return, but to also pay the proper postage on the envelope. This provides the kind of special service the client remembers, and which indicates to him that you are doing professional, quality work. This stamped envelope, of course, enables the client to mail his return immediately after signing it, without having to hunt for stamps, or worry about having sufficient postage.

Such thoughtful consideration on your part can only help your practice to grow—both in quality and in quantity. It's a small investment that can bring a handsome return.

Chapter 8
STREAMLINING YOUR OFFICE PROCEDURES

The art of personal efficiency...involves weaving the cables of constructive habit so that right action will become automatic. In sport and in business—good habits mark the champion.

Wilferd A. Peterson

Time is a man's most precious possession—his most precious commodity. To take a man's time is to take a portion of his life. To give a man some of your time, is to give him a portion of yours.

Margaret E. Mulac

Many practitioners believe that their sole concern should be to attract as many clients as possible. They feel that once they have a steady clientele their problems are over, and the rest will somehow take care of itself. But this is far from the truth; for unless you institute effective office procedures at the very start, you will find yourself constantly engaged in an uphill struggle. As your workload increases, you will find yourself steadily overwhelmed with paperwork; so the time to plan for an orderly, manageable flow of work—in and through your office—is at the very beginning, before it ever becomes a problem.

Your Office Filing System

Every office has a habit of generating innumerable pieces of paper, and filing is a process for organizing and storing such information. It is often referred to as 'the memory of any business'; therefore—readily retrieving and replacing information is an essential key to effective office procedure. A good filing system, at the heart of your practice, shows that you are well on the way to operating a smooth running tax facility.

Speaking in a broad sense, filing procedures include—

1. Your client files.

2. Your tax forms, schedules, and related office supplies.
3. Personal files, and reference material.

Client files, in most offices, are maintained in alphabetical order by a client's last name; though some of the larger firms using computerized services are converting to an account numbering system. However, for most purposes, you will find the alphabetical system to be the fastest and easiest to use. We suggest you use a legal size folder for each client in which you keep all copies of his tax returns, related worksheets and papers, memorandums, and everything else pertaining to his case. Wherever you can manage to clip or staple items together will make working with the file even easier.

In the case of a business, if it is not incorporated, the file should be kept under the client's name, if a corporation, file it under the corporate name. Partnership papers should be filed under the partnership trade name. A great time saver is to mark the client's social security number or business ID number prominently on the outside of the file. This not only guards against mix-ups where clients have similar names, but also gives you access to the number at a glance—because very often all you need is the client's number.

It is important to keep filing procedures as simple as possible. This may not be of such vital importance as long as you run a one-man office and have everything at your fingertips, but as you add office assistants, a complicated filing procedure can become hopelessly confusing. Therefore, it is important to be consistent throughout, whatever file system you choose for client information—use that system exclusively. Don't inter-mix name, subject, or number files within one system.

Generally—Federal, State, and City tax returns should be filed together, unless the papers are so bulky that separate folders are required. Even in that case, all tax returns for each year should be kept together.

As for the sequence of material inside a general folder, it can either follow a 'date' order or an 'alphabetical' order—whichever you find more useful. But again, be consistent in whichever you choose.

Some practitioners use color coded files: the regular manila for individual returns, and various colored ones for different types of business entities. Color coded files are also helpful in preventing and locating misfiles.

Another handy type of folder to use is the one with build-in fasteners, to assure you that no papers get lost or misplaced. When using file folders with build-in fasteners, some use the right hand fastener for all current material (i.e. items still being processed), and the left hand fastener for all completed work (copies of all returns, and so on). Eventually, as the left hand side fills, the material is transferred to storage files.

As we mentioned earlier, it is imperative that you keep all worksheets, notations, and other pertinent supporting material in your client's file. We also recommend that you indicate on your file copy of the return—all supplementary schedules; or on your working papers—the nature of substantiating evidence (or an explanation of how an estimate was arrived at) available to support the claimed deduction, in the event of an audit. Although only a small percentage of returns are audited, there is always the chance that any particular return will be the one so honored. There is no better way to prove your competence, thoroughness, and foresight—to both client and IRS examiners—than by having this vital information at your fingertips, if and when the need arises. (More information in preparing for and handling IRS or state tax examinations will be found in a separate chapter later).

If, with time, a client file gets too thick and overburdened, break it down into subfiles; create another folder to avoid important papers getting crushed or becoming illegible.

A good filing system should also include a schedule of finally discarding unnecessary records, or of at least relegating such material to an inactive file. With this in mind, try to keep all records, including your working papers, for at least six years. In most cases, however, records can be moved out of the active files after two or three years, when the likelihood of an audit becomes remote. If file cabinets becomes overcrowded, you can obtain inexpensive fiber board transfer files for inactive records. These storage files can be stacked or stored on shelves, at a considerable savings in the cost of filing equipment.

Make sure to return to every client all material that is not required in your file. First of all, this prevents your files from becoming bulky and clogged. And secondly, it eliminates many frantic calls from clients in search of a particular document. There is always the chance that an important paper may get lost or misplaced in your office. To protect yourself—always put a note on your client's records stating what was returned to the client along with when it was returned.

Note to computerized preparers: Suppose you have all your returns on computer, should you still produce and retain hard-paper copies? What about supporting documents, worksheets, notations, etc.? Most practitioners will agree that a file copy of every return prepared should be printed out. When a question arises about the return or any particular item it is much easier and quicker to access and review it on paper than on screen, where you may have to switch back and forth between different screens.

As far as long-term retention of return copies is concerned, the jury is still out. Some destroy them after a year or two when the likelihood of their being needed becomes more remote. Others hold on to them indefinitely. Our feeling is that if you're short on filing space and you have a truly reliable backup system in place for your computer files you can probably safely discard the returns after a year or so. Otherwise don't touch them. We've heard too many horror stories about crucial files inexplicably disappearing into cyberspace.

Supporting documents, worksheets and other pertinent papers should be filed away and retained for as long as they may be relevant.

A Few Filing Tips

1. Straight cut folders, where the exposed tab runs the entire length of the folder, are the most versatile; and can be used in shelf or cabinet filing. Any size label can be used (and can contain various information) because the tab is so large.
2. Make sure you label properly and legibly, so others can also readily refer to the files; and be sure to label file drawers properly as well.
3. You may find that a separate personal file for your own documents and records will work more effectively if you use an alphabetical system—by subject (i.e. Advertising ... Bills ... Car ... Checking Acct ... Entertainment ... Insurance... Medical...Office supplies...Speeches...etc.—the list goes on, each practitioner according to his needs).
4. Try to designate a specific time of day (or week) to sort, and either discard unneeded papers or file them into their proper location.

Time Management: Making Good Use of Your Available Time

We all have the same amount of time each day to work with. The differences come in, however, in how each of us handles this precious resource. And although some seem to make better use of their time at work than others, there are principles and techniques that you can master to help improve the way you handle your time. We will discuss some of these techniques in this section, and it will be worth your while to give them a try, for here is a skill that bears directly on your professional progress.

A Time Log

Keeping a temporary time log will allow you to see how you are currently spending your time. For a week or two, try to keep track of what you are doing: list your work activities and record how long each activity took you to do. Also make note of interruptions, and how long they took, and if they were necessary—or if they could have been handled in some other way.

Keeping such a time log of your work habits will give you a reasonable picture of how you are presently handling your work day. After a week or two examine your records: study them, evaluate them, and see where there are areas that can stand some measure of improvement. Then, when you try out any time management techniques, you will be in a better position to apply them to specific areas of weakness.

Planning Ahead

The technique that we will more closely examine we will simply call 'planning ahead', and we will break the material down into three parts: listing activities, setting priorities, and scheduling.

Bear in mind that the problem with the use of time is not what to do, we usually have enough things to do to occupy our time. The problem is one of how to get the jobs done. And toward this end proper 'planning ahead' can be of great help. Setting aside some measure of time to plan now can save you hours of confusion later.

1. **Listing activities:** Plan your day in advance by beginning your work day with a written list of all you intend to take care of that day. Depending on the type of person you are, you may find it easier to do this early in the morning before the work day begins, or perhaps on the previous night, before you retire. But do not get into the habit of devising this list once you get to your office or work area. This particular technique is more effective when you arrive at work with the list complete, and you already knowing exactly what you have to do.

In this regard, planning ahead serves you like a blueprint serves a builder: once your plan is on paper—laid out in front of you like a blueprint—it becomes your challenge to structure the day in a way that best approximates that design. Of course, all of this is easier said than done, but having it in writing, in the form of a plan to carry out, provides you with direction, a pattern, a formula if you will. Having your plan in writing also makes the tasks more concrete and tangible, in a way it is as if they are somewhere along the way toward completion—you've already started them, and starting is sometimes the hardest part.

Another benefit of devising lists is that you will develop the good habit of 'viewing' the day ahead. Later, as you raise your sights and develop a longer range view, you will better envision the week ahead, the month ahead, and even the year ahead. In conjunction with this you can make use of any number of helpful calendars—from pocket size to wall size—depending on what serves you best. But on these convenient calendars that provide you with 'a week at a glance', or 'a month at a glance', you can chart or fill in the various tasks you know you have to do.

Developing this kind of vision is a beneficial by-product that derives from the power of planning ahead.

2. **Setting priorities:** This step is really a refinement of the list you made in the previous section. Obviously, not all the things you may have listed are of equal importance. Some items may be pressing and carry a high level of priority, while others may be pushed off for a day—or more, though it might be nice to take care of them today. But to determine which items fall into which category, you have to go over your list and prioritize—to indicate high, medium, or low levels of priority. Then you know which items must be taken care of first and which can follow later.

3. **Scheduling:** Now that you have listed and prioritized your daily (or weekly) tasks, you still have to schedule the tasks into your time of day.

   To schedule a task is to determine when to do it, and how much time to give it. And of course, with time and experience you will always be better able to judge how much time to allot for certain activities. You will also get to know your work habits, and assess when in the day you accomplish the most and when you accomplish the least, for all of these considerations will affect the way you finally schedule certain tasks.

   You will also quickly learn that you cannot control the entire day. Hardly anything works out exactly as we planned it, so you will have to factor 'flexibility' into your schedule. You will have to leave some room for the unexpected, slots of in-between time to take care of certain 'unforseeables'. And this also includes some quiet time for yourself—to rest, return calls, respond to the mail, or catch up on other unfinished chores.

   Planning ahead can be a helpful technique in streamlining your office procedure, and like any good habit, it will involve some initial investment of time to properly cultivate. However, once you have mastered the system, you will see that making the list will only take a matter of minutes each day, but these are minutes that will literally save you hours.

   There are a number of excellent software programs available to help with scheduling (both tasks and appointments) and time management, create to-do lists, keep track of unfinished projects and remind you of upcoming due dates and deadlines. But, unfortunately they require the same amount of self-discipline and dedication as traditional methods. Unless you put in the necessary time and effort to make the required entries and keep them current, the most sophisticated program will be useless.

Tax Preparation and Efficient Use of Time

Here are a number of suggestions to help you and your staff make effective use of time—especially to avoid peak season bottlenecks.

- Make sure you have all the required papers: W-2 Forms, 1099's, etc.; as well as all supplementary data such as rental income, expenses, securities transactions, and so on. This will help you avoid the common practice of interrupting work on one return because vital information is missing; switching to another return, and then having to go back to the first one. This is a time-consuming routine, because every time you switch from one client's paperwork to another's, you have to stop and reorient yourself.

- Try to anticipate 'peak season' workloads by doing as much work as possible beforehand. Some practitioners even prepare tax return kits for each client before the season begins. In this kit they assemble all anticipated tax forms the client will need, and other worksheets or information that the previous record shows will be required. Some go even further and 'head up' their worksheets and schedules with the client's name, social security number, and taxable year—in order to speed work along during the tax season. You can also prepare carry forward schedules in advance, for capital losses, charitable contributions, and investment credit carryovers.
- Avoid wasting time searching through files by making sure that all related records are stored together.
- Make a preliminary list of all forms and additional information you will need for a particular return. This too, will help avoid annoying interruptions while the information is gathered or forms obtained.
- Here is an idea to help ensure that you have all the necessary forms on hand. Every time you use one of the less common forms, make a note of it on a special list. Before the next tax season pull out that list: if you retained these clients, you will most likely need the same forms again, so with a list as a guide, order sufficient quantities of the various forms you need.
- Some practitioners find that they increase efficiency by evenly dividing their day, setting aside one part for client interviews, and using the other part to work—without interruptions—preparing returns.

Time Management: A few practical tips

1. It's part of human nature to drift toward easier tasks, putting off until later the more difficult, problematic assignments. But this usually tends to aggravate the work situation. It's psychologically more advantageous to deal with a difficult matter, you then tend to feel so good that you become more effective at everything else you do that day. With the burden lifted, you are more efficient in the way you handle your other jobs.
2. Be organized: Here is one vital area where time and space intersect. By knowing where things are, you will be amazed at how much time you will save, not having to hunt them down.

3. Unnecessary interruptions can dilute your efforts and force of concentration, so you have to develop a method of handling interruptions. If you have an assistant, then that person can screen calls or take messages for you. If you work alone, then you can leave a polite message on an answering machine and employ that at certain hours of the day. But you must develop a strategy that enables you to devote certain blocks of time to uninterrupted work.

4. To overcome procrastination, try to establish deadlines for yourself—for both beginning and ending a task. Also, reward yourself and others (i.e. family members) at certain points of progress or at the completion of a task.

5. In dealing with mail the goal is to handle each piece of correspondence one time only. If at all possible, do what has to be done right away, rather than reading it, putting it down, and then having to start over with it at some later point in time.

   When you open a piece of mail, try to ascertain immediately where it belongs: whether in the circular file or some other appropriate bin or file. For example, put all checks away where they belong for later deposit.

   If it something you can handle on the spot—with a brief note or telephone call—then do so. If you can't, try to develop the habit of putting everything in its proper slot immediately—to be taken care of in its appropriate time, rather than repeating the cycle of having to senselessly read and re-read your mail; thereby wasting precious time.

6. Economize your movements: If you have errands to run or traveling to do, try to plan your day so that you can take care of as many stops as you can on a single outing, whether it's trips to the post office or to the bank or to purchase supplies.

   And if you happen to be waiting somewhere for a period of time—for example, in a dentist's or in a client's office—you may find that there are certain tasks you can take care of while you wait: sorting, figuring, bills, brief notes, scheduling, or composing a letter. Waiting time can be put to creative use to handle jobs that might otherwise cut into prime working time.

What Records to Keep

- You should keep a master record (a 4" x 6" index card works well) for each client listing his name, address, telephone number, date you prepared his return, form of return, amount of the fee, and date of payment. These cards will be very valuable to you the following year. You may want to send reminders or post cards to this list, or you may want to advise them that you moved or have a new telephone number. You will also be able to tell at a glance how much you previously charged each client.

- You may also want to make special notations on some of the cards for future reference. For example: 'Wasted time because of incomplete records'; 'Ask client to bring more detailed lists'; 'Complicated return, increase fee next time'; 'Recommended some new clients', and so forth. You will also find this file to be a veritable gold mine in case you want to sell or merge your practice. These cards can either be attached to the client folders or filed in alphabetical order in a separate card file.

- One form many practitioners and tax firms use is a Master Control Sheet on which they list each client or return, showing the initial interview date or first contact with the client, any follow ups required, date return was completed, fees, etc. This gives you an immediate profile of each client. As your practice grows, such a form can be quite helpful.

- We also suggest the use of a calendar on which to note the date each client was interviewed in previous years. This enables you to follow up on a client if you haven't heard from him at about the same time in the current year. It's a great help at bringing in clients, and eliminating that last minute rush.

- As a tax practitioner you will be intimately involved in the bookkeeping of others; at the same time you must realize that it's imperative to keep accurate records of your own if your business is to succeed.

  Aside from providing you with your own income tax information, maintaining accurate records offers you valuable information that can help you see developing trends in your practice. Examining and comparing your records from year to year will reveal to you your practice's strengths and weaknesses, and armed with such vital information you can better chart the future of your practice.

  So for your own income and expense record, be sure to get a good, easy-to-use cashbook. Simply list all receipts for each day with the names of the clients, along with expenditures and brief explanatory notes—if necessary. At the end of the day or week, transfer the totals to a separate 'Summary' page. It's also a good idea to do your own bookkeeping regularly, so that paperwork and confusion do not mount up.

Keep Records of Communications with Clients

In the course of your practice you will make and receive innumerable phone calls—to, from, and on behalf of—clients. It is highly recommended that you keep a record of these calls—especially the more important ones. Some clients require an inordinate amount of time on the phone, and unless you keep a record, you will never be compensated for this time. Furthermore, you will receive a lot of information over the phone, and unless you record it immediately, you may forget it. In addition to causing you embarrassment, you may have to do much of the work over again. If you choose then to write such information down, it may be worth your while to keep a telephone log. A phone log is charted paper that allows you to keep a careful track of who called, when they called, the length and reason for the call, what resulted from the call, and you can also make note of any long distance charges to get reimbursed—if that's appropriate. On the other hand, some practitioners use a tape recorder or attach a recording device to the phone to record such information, in that case you should advise the client that the conversation is being taped.

Some practitioners, like doctors, establish special telephone hours during which time they do business over the phone. At other times calls are answered by the secretary or an answering machine which takes the information or assures the client that the practitioner will return the call as soon as possible. This avoids interruptions and also enables you to better serve the calling client by pulling out his file and looking over his papers to refresh your memory, before returning his call.

Complete All Work As Soon As Possible

One of the more common human shortcomings that may interfere with the efficient functioning of your tax practice is the tendency to procrastinate—to put off completing a return until the last moment.

When a return is prepared, let's say, in February, and a question arises requiring further research, or you find yourself in need of more information; you then have plenty of time to do the research or obtain the missing data. But on the other hand, if you are closing in on an April 15th deadline, or are in the midst of the pre-deadline crunch, you'll be tempted to skip or abbreviate the necessary research or even prepare the return on the basis of incomplete information.

Now if a client, in spite of your urging, insists on bringing you his material at the last moment, he has only himself to blame if he does not receive the top-flight service you are prepared to give him. But the client who comes to you before the rush starts, who makes it a point to assemble all relevant information, and documents early in the season, he has every right to insist that you give his return the most thorough and painstaking attention possible. He deserves no less!

Hand in hand with the tendency of putting off a return until the last moment is the habit of many to begin working on a return and then leave it incomplete for days or weeks at a time. What happens all too often is that, to conserve time, the tax preparer takes down basic information and then puts the whole folder or file away until he has a more opportune moment to complete the return. However, when he finally does get around to it, he must go over all the information, reorient himself with the client's facts and circumstances, and try to recall what was said or done until now. Not only is this wasteful and inefficient, but you invite the costly risk of overlooking or forgetting vital facts and information.

In short: the best time to prepare a return is right away, or as soon as possible after the interview with the client, or after you receive the necessary information. If you must interrupt because some facts are missing, make every effort to obtain the information as quickly as possible, so you can proceed with the return while all the information is still fresh in your mind.

Getting the Most Out of the Telephone

The phone with its incessant ringing and disregard your schedule can be both friend and foe. Employed judiciously, it is a tremendous timesaver and goodwill builder, without proper safeguards it can be a disruptive and annoying intruder.

First of all, make sure that you—or whoever else answers the phone in your office—cultivates a pleasant telephone personality. Remember that the phone is an extension of yourself and of your office. Clients and prospective clients contact you more often by phone than in person, which means that it can create impressive amounts of good- or ill will. No matter how busy you or your secretary are, and regardless of how ill-mannered or unreasonable the person is on the other end of the line, every call must be treated graciously and politely.

At the same time, you owe your clients your full attention while you are with them. If a call comes in during a conference or interview, try to keep the call as brief as possible. Better yet, tell the caller you will return to him as soon as you are free—and make it a point to do so. It can be annoying to a client to have his conversation with you constantly interrupted by phone calls. Even worse—he may suspect you are charging him for the time you are spending on the phone.

If you have a secretary, let her screen your calls and explain tactfully that you are presently engaged and will call back shortly. Likewise, if you have to call a client on a matter that may take more than a few moments, ask him first whether he has the time to talk now, or whether you should call him back later.

Telephone Courtesy

Very often, the telephone is the first—and only—way a potential client has of judging you and your practice, so be sure to employ telephone courtesy at all times.

Also bear in mind that the telephone calls for communication skills that differ somewhat from those used in face-to-face contact. With a telephone you must convey your attitude—your sincerity, and sense of confidence—through your voice. Here, then, are a few points to consider in order that you can make wise use of this vital instrument.

1. During business hours, the telephone should be answered—by yourself, a secretary, or an answering machine—every time it rings. While on the phone, be pleasant but businesslike, relaxed and sincere.
2. A business conversation should be free of all background noise—even music.
3. When preparing for a call try to have all pertinent information at your fingertips: client files, calendar, appointment book, etc.
4. Consider carefully how you answer the phone. Do you just say "hello", and then wait for the caller to respond. This may suffice in your home, but for business purposes it is not enough. You must identify yourself. Begin with an appropriate greeting—good morning or good afternoon—and then state your name and/or your company's name.
5. Don't leave a caller 'on hold' for more than half a minute, if you can help it. If necessary, arrange to call the person back, but don't leave a caller indefinitely and without a reason. It's irritating, and it may even cost you a client.
6. When giving a caller pertinent facts and figures, speak slowly and distinctly—and speak into the telephone. Many are in the habit of cradling the phone in such a way that it juts into the speaker's neck or chin. This muffles the sound and can easily frustrate the caller.
7. Be aware of the importance of language. Over the phone a client only has your words to go by, so be careful of both what you say and how you say it.
8. Listen carefully and effectively. Listen to what is being said and how it's being said, and don't interrupt the caller unnecessarily. You can even reinforce the client by acknowledging your understanding of his remarks in some way.
9. Don't hang up on your clients. Wait until the other party hangs up and then proceed to do the same.
10. In terms of both time and money, the telephone is your most effective communication tool: exceptional for scheduling appointments, sharing information, answering questions, and settling matters. Learn to use it well and use it wisely.
11. A final thought. 'Call waiting', while heavily touted by the phone companies, can be most annoying to the person whose conversation with you is being interrupted by an intruder. As was already mentioned, the party on the other end of the line deserves your full attention without the distraction caused by continuous, impatient, beeps.

Business Correspondence

Handling correspondence is an important part of regular office procedure. You will often be called upon to communicate your thoughts in writing—either with letters you initiate or letters you respond to—so this is an essential skill, the mastery of which, will help you immensely in the smooth running of your practice.

Remember, too, that the letters you send—their style, content, and appearance—are extensions of your office personality. They are your thoughts and voice enclosed in an envelope, and in some cases, the impression a letter makes, can spell the loss or gain of an important client.

Here then are some practical thoughts and guidelines to help you write more effective business letters.

1. You can write a good business letter and still sound natural and relaxed. You can deliver your message in a businesslike manner, without sounding stiff.

   For example: Instead of writing 'according to our records', use 'I received your letter'. Instead of 'due to the fact that', use 'because'. Instead of 'in the amount of', use 'for'. Instead of 'at the present time', use 'now'.

2. Be pleasant and positive in your writing. Sprinkle the letter with courtesies. Effective correspondence can go a long way to strengthen a business relationship. Instead of 'We have received your recent forms', use 'Thank you for the forms you sent'.

   At the same time, be tactful and diplomatic when relaying unpleasant messages.

3. Be personal: use the pronouns I, we, and you, whenever appropriate.

4. Before you start to write, be clear about the purpose of your letter: what do you want the reader to know or do?

   Spend a few minutes thinking about the purpose, jot down some notes or make an outline; then turn it into a rough draft. The final letter will then be much easier to write. Obviously, a word processor makes letter writing even easier—almost painless.

5. Once you have written the letter, be sure to check it for accuracy. Review all the facts and figures to make sure they are correct, and proofread the letter for mistakes—in names, spelling, and grammar.

6. Remember: although style, content, and form are important, the real effect of your letter depends upon the words you choose to carry your message. Keep your words clear, simple, and to the point. Write in a friendly tone; businesslike, but not overly formal.

7. Surround the letter with balanced, ample margins, and try to center your letter on the page. If your letter is brief, don't put it all high on the page. If your letter is long, don't cram it all onto one page; use an additional sheet of paper—even for one paragraph.

8. In business replies, try to be prompt; don't make others remind you repeatedly of their request.

A parting thought to this chapter

Maintaining efficient office procedure calls for the coordination of various talents and skills. At first you may wonder how all these components will ever fall into place; but with time and experience it becomes one unified effort—harmonious, second nature, almost automatic. With office procedure it's exactly the same: by giving it time, practice, and proper attention, you develop the habits, traits, and skills, that make managing an office as routine as...driving a car.

And just like your car—once your office is in gear, and you know how to guide your vehicle—you'll be advancing with certainty on the road to success.

Chapter 9
BUILDING YOUR TEAM

A leader does not say, "Get going!" Instead he says, "Let's go!" and leads the way.

He helps those under him to grow 'big' because he realizes that the more 'big' people an organization has, the stronger it will be.

A leader does not hold people down, he lifts them up.

A leader has faith in people. He believes in them, trusts them, and so draws out the best in them.

A leader uses his heart as well as his head. He is not only a boss, he is also a friend.

He has a sense of humor. He is not a stuffed shirt. He can laugh at himself. He has a humble spirit.

A leader can lead. He is not interested in having his own way, but in finding the best way. He has an open mind.

A leader keeps his eyes on high goals. He strives to make the efforts of his followers and himself contribute to the enrichment of personality, the achievement of more abundant living for all, and the improvement of civilization.

William F. Pederson

Hiring Staff Members

Even if you are still months away, it pays to give thought and take practical steps toward achieving maximum efficiency and productivity during the busy tax season. And of all the labor-saving devices a modern office can have, the most valuable are those that extend your precious time.

In this chapter we will discuss a number of ideas designed to give you—the Tax Professional—more time to perform the tasks that only you can do best; relegating all other important—but tedious and routine work—to others. In hiring office help, you want to find someone who can take care of paperwork and other details—leaving you to ply your craft and to dealing personally with your clientele.

Regardless of how many clients you serve, all necessary tax office work can be classified as either:

A. Routine clerical tasks, or -

B. Skilled, technical, professional work.

It follows then, that all or most of the tasks in the first category can and should be delegated to others. This enables you to concentrate on the work that requires your specialized training and abilities.

Using an Office Assistant

Even the smaller tax practitioner can employ—to good advantage—the services of an assistant to perform the following tasks:

- Act as receptionist while you interview or prepare the return of another client.
- Make appointments for you. Most appointments come by phone, and by using an appointment book your receptionist should have no trouble setting up and keeping track of your appointments. Just make sure the receptionist records and double-checks the information so that it is correct, legible and there are no scheduling conflicts.
- Take care of many vital but routine office tasks such as addressing envelopes, collating complete returns into sets, inserting and mailing, billing clients, etc.
- Answer routine questions. Many of the calls a tax consultant receives (especially in those hectic weeks prior to April 15th) can easily be answered by an informed assistant: Where do I send my federal or state return? When will I get my refund? How much tax do I have to pay? How do I make out the check? Such calls are interruptive, and they eat up a lot of time, especially if they come in the middle of an important discussion or interview. Having someone else field such calls will enable you to give your undivided attention to your clients, and it will earn you the respect of the caller as well.
- Follow-up calls. Usually, there are a number of clients whose returns cannot be completed immediately, they may be lacking information or documents that the client did not bring along. It is most advisable that you call the client after a day or two, if the information did not arrive. A simple 'tickler' file, which the receptionist is in charge of, can accomplish this.
- Have your assistant prepare all necessary copies of forms and other documents.

- If your receptionist can type, let her prepare all supporting documents—schedules, explanations, statements, and other correspondence.
- Train your assistant to handle the files—preparing the file folder for a new client, filing all completed copies, and retrieving client copies of previous returns in advance of each scheduled appointment.

Each of these tasks considered alone may only amount to a matter of minutes, but you will soon realize that if you can eliminate 10-15 minutes worth of interruptions per hour with a client, the quality of your service will greatly improve. You will then be able to give your client full attention and the concentration he deserves.

Your first office assistant will probably work only part time and even then, only during the tax season. Your helper may be a spouse, another family member, a high school or college student, or even a full time secretary in need of additional income. But always make sure to hire a responsible, mature, intelligent person. Then, the first thing to impress upon any prospective employee is that all client matters are strictly confidential. Gossip, about any client—in or out of the office—is absolutely forbidden.

Next, take time off to train the employee in all the duties you expect her to perform and then delegate as much work as possible to her. Let her handle all filing, retrieval of files, appointments, and so on. Bear in mind that the longer the individual works in your office—assuming the work is satisfactory—the more valuable he or she becomes as a member of your organization. If you encourage her to take on more responsibility, she will be able to relieve you of more routine tasks. At the same time, clients will get to know and trust her as well. Therefore, do everything you can to hold on to a good, trained and trustworthy employee

If the extra help you need is, in fact, just to get you through the tax season, then you might also consider hiring someone through a temporary employment agency. The understanding here is that the work is of a temporary nature—there are no employee benefits or other peripheral issues to consider, and you get a qualified helper without having to go through the procedure of advertising or interviewing. Furthermore, the agency is seeking to please its clients, so if a particular individual is not working out well, you can simply call for a replacement.

To get your money's worth from a temporary employment agency, be as precise as you can in describing the work you want done. Reliable agencies have extensive files and will do their best to find you a worker with whom you will be satisfied. In fact, it often occurs that a 'temp' becomes thoroughly familiar with the procedures in a particular office setting—and if the 'chemistry' is right—this individual goes on to become a permanent employee.

Full Time or Professional Employees

As your practice grows and devėlops, the quantity of work you must perform expands as well. When it reaches the point where you need more than just clerical assistance and you'll need to consider employing a qualified tax assistant, or at least train someone to handle the less involved tax returns.

If you prefer a trained, experienced professional, you may locate a retired accountant or other tax practitioner willing to do part time work. For leads and contacts, we suggest that you ask other local practitioners, or put a small ad in your local newspaper.

The role of a professional assistant is to interview clients, and to prepare the simpler—or even more complex returns—depending on his or her competence. It is also important to choose an individual who is not only technically competent, but who can relate to and properly deal with your clients. At the same time, be sure this individual is well trained and up-to-date on all new tax developments.

If you happen to come upon the right person, you may find that in the long run it is more economical and practical to train that assistant yourself—provided you have the necessary patience and instructional skills to do so.

Whichever course you pursue, you should realize that the hiring of full-time professional help (or helpers) marks a genuine turning point in the development of your practice, and it is therefore a matter you should consider with care. An employee, though he or she may differ from you in significant ways, is nevertheless an extension of yourself and your tax practice. Your clients' attitudes toward your practice will be strongly influenced—favorably or unfavorably—by their contact with your assistants, so hiring someone should not be rash, desperate, or a spontaneous gesture. It calls for a definite approach or plan of action.

Here, then, is a sequence of steps you can follow to help you locate and recognize good, prospective office assistants.

Writing A Job Description

It isn't enough to have a vague idea of what you want hired help to do. The more specific you are, the better off everyone will be. So first try to identify your needs and define the tasks you want someone else to do. For example (when seeking clerical help)—typing, billing, handling mail, filing, handling incoming calls, writing letters, making appointments, general bookkeeping, etc. And if you can be more specific within these categories, that's even better.

It would then be a good idea to prioritize these tasks. Which are primary duties and which are secondary or even expendable duties? You may not find someone to handle all of the jobs on your list, but if you can get someone to capably handle the most important ones, that may be good enough.

Then write a brief job description—including job qualifications you require—spelling out the duties, responsibilities, conditions, and requirements that constitute the offer you are proposing. This precise description will be a helpful guide to you and job candidates as you proceed through the hiring process. It will provide everyone with a clarity of what the job entails, and will help to avoid later misunderstandings. You can also use this description as a basis for your want-ads or the information you transmit to employment agencies.

Recruitment

You are now in a position to recruit applicants, and there are a number of employee resources available to you: secretarial agencies, employment agencies, government agencies, educational facilities (for example, schools that train the kind of worker you want), putting a 'Help Wanted' ad in your local newspaper, or posting notices on select bulletin boards, and you can also send the word out among friends, relatives, other practitioners, and other employees.

Whichever avenue you pursue, you will have to be ready to receive prospects—either with a filled-out application form that records basic and essential information, or with a resume. When you sift through such material you can already begin screening out acceptable from unacceptable applicants. Carefully examine the filled out forms, questionnaires, or resumes: Is the information neat or sloppy? Is it clearly written or incoherent? How is the grammar and spelling? Is the writing 'to the point' and functional—highlighting skills, experience, and abilities; or does it 'beat around the bush' and speak about secondary concerns and hobbies? Does the writer's tone reflect a positive, serious, and ambitious note, or is it merely casual and flippant?

Interviewing An Applicant

Even after you are left with your most impressive applications, you cannot hire someone without conducting a well planned interview. You need to know how applicants present themselves physically, and how they conduct themselves in conversation.

- You can prepare for your interview by writing out appropriate questions in advance. Such questions might include: What were your responsibilities in your last job? What did you like or not like about your job? What did you like or not like about your boss? Why did you leave? Aim for questions that are 'telling', that go beyond a simple response of 'yes' or 'no'. This way you can get a feel for the way this person thinks and looks at things. Ask questions that go into details and specifics—that touch upon skills and situational encounters.

- Conduct the interview privately. Most applicants are somewhat nervous at such encounters, so you may try to put the person at ease by offering a drink, or by first engaging in some light conversation. In such a setting the candidate's replies will tend to be more candid and meaningful.

- Aside from the questions you ask you should also make observations about the person's style of presentation—his or her manner or dress, and discreetly jot down information you will later want to consider.

- At the end of the interview, graciously thank the applicant, and inform him that you will soon be in touch. When you do call back—with one answer or another—thank the applicant again, and briefly explain the decision you came to.

  However, (before you give any reply) make sure you contact references—both listed and unlisted—and ask former employers very specific questions: Was the person punctual? Polite? Efficient? Would you hire this individual again?

- Some, in the course of an interview, will give tests to applicants of various job related skills, to get an even better gauge of a candidate's performance levels.

- Although it is difficult to measure certain important attributes in an interview (i.e. motivation), this is still an important and necessary step in seeking a reliable, prospective employee.

Selecting The Best Applicant

It may seem obvious, yet it's important to point out that you are primarily looking for someone who can do the jobs you need to have done. Don't pick candidates just for peripheral reasons—a winning smile, good manners, or a way with words.

Also; don't look for a mirror-image of yourself, or for someone who handles everything just as you do. Different personalities lend themselves to different kinds of work, with various likes and dislikes, and tasks that you find tedious, another will find challenging. Of course there should be harmony and compatibility, but you are not searching for a carbon-copy of your attitudes and skills. You want people who have the drive and the ability to get your work done.

At this point take your top-choice applicants and consider your material thoroughly: written forms, phone conversations, interviews, and references. The closer you come to fitting the applicant exactly to the job, the happier things will be for everyone concerned.

You should make your selection carefully, but also do it as soon as you can: good candidates have probably applied for other positions as well, and may get hired before you make up your mind.

Finally, don't throw away any of the prospect files you were seriously considering. They may be of use to you later on, if a present employee does not work out and leaves you, or if your practice expands and you need more help—these files may come in handy.

Training and Orientation

Misunderstandings between an employer and an employee can cause resentment, office tension, and loss of productivity. One excellent way to reduce such misunderstanding is to take a new employee and clear the air of all questions and concerns beforehand. Either in writing or orally, make sure the employee understands what he or she will be expected to do, and what you will provide in return for that service.

Among the points to be covered are the following...

- a list of job duties.
- date the job will begin.
- the amount of pay for the work, stated in terms or pay per hour, week, month or year.
- the day of the week or month the employee will be paid.
- fringe benefits, if any.
- whether there will be paid vacations, and length of vacations; sick days, 'personal days' and; holidays policies; maximum number of paid sick days, and a list of holidays.
- time of reporting to and leaving from work, and overtime policy.
- attendance and emergency leave policies.
- work rules and dress code.
- lunch time, break time, and personal calls policy.
- at what intervals the employee will be considered for a pay raise.
- discipline and termination policies.

After hiring and orientation you still have to train your new employee. This is really an ongoing procedure, but there is still a 'breaking in' period of on-the-job learning-by-doing.

Here are some guidelines that can help you through this initial stage:

- Prepare the employee by first putting him at ease. Be patient and encouraging. Find out what he already knows and then proceed to tell him precisely what you are going to teach him.
- Present the new task in steps—one step at a time—showing the employee what has to be done, never demanding more than he can master at one time.
- Supervise as the employee explains and shows you his own performance of that task. It's altogether natural to make mistakes, just correct the employee's errors and periodically check on his work.
- To help motivate an employee, try to delegate tasks that encompass greater degrees of responsibility...to oversee a job from start to finish, rather than giving out pieces of repetitive work. This makes the job more interesting, complete, and satisfying to do.
- With time, you have to give your workers enough authority to carry out tasks on their own. This also offers them greater freedom to err, so you will have to periodically check on their progress. But it's counterproductive to stifle your workers, if you have good, capable workers—trust them, and employ their skills.
- Encourage your professional staff members, or trainees, to read tax related periodicals, announcements and news releases in order to keep abreast of information that pertains to your work; and if you have the time—discuss as many client problems and returns with them as possible—exposing them to a wide variety of tax situations.
- For the first six months to a year under your supervision, if you see that your employee is capable, you can advance him to more responsible levels of tax work. Your staff member will also gain a lot of valuable experience if you let him do the research on the tax problems you are working on. Just make sure you check the work until you are certain that you can fully rely on his findings.
- One question to resolve is whether to let the trainee do his studying and research work on his own time or on your time. The general trend is to permit studying on employer time, to encourage as much independent study as possible.
- As your relationship with employees develops, remember: Everyone is different in significant ways—in goals, attitudes, and personalities, and that no one is perfect. Everyone is subject to moods, ups-and-downs, and unexpected problems—all of which may affect performance on the job. Come to understand this within yourself, by way of your own strengths and weaknesses, and then your understanding, tolerance, and treatment of others will go much easier.

Working With And Supervising Others

At this point, with one or more people working under you—you have technically become a 'manager'. And managing others—even one other person—adds a significant change to your base of operations. For one thing, being a manager may call for interpersonal skills and abilities that you haven't had any reason to cultivate until now. After all, you were trained to operate an effective and efficient tax practice, and now—as your practice grows—you are being asked to properly supervise those who are under your hire.

Here then is a set of guidelines that will help you in your capacity to supervise anyone who works for you.

- As the employer you should set a good example. Show up on time and dress appropriately. Your own conduct, appearance, and attitude will set the tone for your staff. Be a model of what you expect and your employees will pattern themselves after you.
- Maintain a healthy, positive, productive outlook, and let your actions and words convey this message to those who work for you.
- Office morale can be affected positively or adversely by an incident that may seem quite insignificant to you, but which is very important to your employees. Don't overlook this factor, and try to treat all incidents with fairness and equity.
- Express orders properly—with decency and respect. A simple 'please' and 'thank-you' are almost always in order. Give clear instructions and deadlines, whenever appropriate.
- Maintain a healthy sense of humor, and—most of all be able to laugh at yourself.
- Don't use obscene or vulgar language, and don't make jokes at an employee's expense. It's demeaning to the employee, yourself, your professional image, and to your practice as a whole.
- The traditional phrase 'firm but fair' is appropriate to your office setting. Be sure your employees know the rules you expect them to work by, and then enforce those rules in a just and consistent manner.
- Be rational, patient, pleasant, and calm. Use self control at all times. Things are bound to go wrong occasionally, even get out of hand, but you must try to remain cool, calm, and in control. This may seem like a tall order to fill, but it's preferable that you remain part of the solution and not become part of the problem.
- Acknowledge an employee's good work. In other words—give credit where

credit is due. Let an employee know that you take note of and are actively aware of—and appreciate—what he or she is doing. In fact, such praise can be lavished openly, even while others are around.

- On the other hand, if criticism is in order, take care of that softly and in private. And when you criticize, make sure it is constructive and not destructive: focus in on a specific action and don't be critical of the person as a whole. Never insult or 'put down' an employee.
- When you make a suggestion or request, give your reason for it, if possible. You will get a lot more cooperation from people if they know what they are doing and why they are doing it.
- Impress upon clients that you have full confidence in your staff. This not only reassures the client, but boosts your office morale as well.
- Don't hesitate to discuss a problem with an employee or ask for his advice if you think he is qualified. This will enhance your assistant's sense of self perception and of loyalty to your practice.
- Let employees use their own methods, as long as they achieve the desired results. The employer who incessantly meddles into every office routine destroys his employees' initiative and incentive.
- Keep communication lines open, and be a good listener. Be available so that employees can come to you with questions and concerns. Encourage workers to discuss problems and offer ideas.
- Express a personal interest in your employees. Let them know that they are not just cogs in an office wheel, but important and valued members of a close-knit team.
- Remember: happy, satisfied employees will do substantially more and better work than unhappy employees.

To sum it up, managing 'human resources'—exhibiting leadership—involves far more than just issuing orders. It calls for interpersonal skills, proficiency in communication, empathy, and a good sense of perspective. You want to direct your employees in such a way that the necessary jobs get done, but in a positive and uplifting—not in a begrudging—way. Toward this end it would be helpful to generally improve your own understanding of human behavior; and in particular—to understand the factors that motivate and energize an employee to do a good job.

The Art of Firing

Like it or not, the time will come when you'll have to lay off an employee. This is never a pleasant experience, whether you have to let the individual go because of downsizing, consolidating, outsourcing, etc., or because of poor job performance, dishonesty, personality problems, or similar reasons.

Moving fast is especially important when dealing with an underperforming or disgruntled employee. A poor worker with an unhealthy attitude and unsatisfactory motivation or performance level can affect the morale of the entire office. So if and when the time comes—and it has to be done—do not postpone the procedure: be prepared to move fast and at the earliest possible opportunity. Putting such things off, and further deliberations can play emotional havoc on your own state of mind, and can undermine your work efforts.

Your feeling unpleasant—even guilty—about it all is a normal reaction, but in your office you will have to keep that reaction to yourself. If you did whatever you reasonably could do for this worker in the way of corrective efforts, and they failed, then it is not your fault that he has to be fired.

The following suggestions may not soften the blow, but they may still be helpful -

- Communicate the message in person—yourself. If you hired the person, then it is your responsibility to relay the notice of termination as well. Do not give this job over to someone else.
- Be prepared. Have all the information you need on hand: the explanations and reasons for the firing, papers, payments—whatever is coming to the employee—so you don't have to drag the ordeal out longer than necessary.
- Be tactful in how you relay the information in order to reduce any tension, bitterness or resentment.
- Where the employee has a satisfactory work history and is being discharged for reasons that are not his fault, declare your willingness to provide recommendations to any prospective employers. Better yet, why not act as a 'matchmaker' by trying to line up a job for the worker—or at least some interviews—with fellow employers. Your concern would certainly be a great morale booster for the entire staff.
- When it's all over, try to assess the situation in terms of how and where things went wrong, so that you can at least try to avoid this sort of circumstance in the future.

Hiring staff members, when it is done with care and an open mind, can be a genuine educational experience. It makes you aware of how complex office procedure can be and of how you can effectively streamline that procedure. Working with staff is a constant lesson in human interaction skills, and it poses a constant challenge as well, for when people master the art of teamwork and cooperation—the accomplishments can be truly inspiring.

Chapter 10
FEES AND BILLING

The subject of fee setting is of great interest to both veteran practitioners as well as to the new start-up. If you're a seasoned tax professional with a well established practice you'll wonder whether you're charging enough, or ought to raise your fees to produce more revenue. If you are a newcomer to the tax profession—or an old-timer who has always been working for others, now venturing forth into your own independent business—setting proper fees is critical to your survival.

- On the one hand you do not want to drive business away by charging too much, on the other hand you have to be adequately compensated for your time and skill. Bear in mind that if you charge too little, the quality of your work will inevitably suffer; in the long run, undercharging will not benefit your clients.

Before deciding on a fee, you must consider a number of factors. The most important ones are discussed here:

a. The average level of fees charged for similar work in your community. This depends to a large extent on the severity of competition, the general economic level, and the size of the community. Fees tend to be higher in better income areas, and are usually somewhat lower in larger cities than in small towns.

As a starter we suggest you find out how much other preparers charge for a simple itemized and non-itemized Form 1040. If you can get a few such figures you will be able to better gauge the fee level in your area.

In many communities you will find several tiers of tax preparers. There are established, professional public accountants CPAs and non CPAs whose major source of income is routine, year 'round accounting work. Many of these individuals or firms expand their offices, hiring additional staff members during the season, to accommodate the many clients who come in only once a year to have their return prepared. The fees generally charged by this group are, as a rule, relatively high with CPAs at the upper end and noncertified accountants somewhat below. They consider tax work a side line though a profitable one, and can therefore assume a "take it or leave it" attitude toward the client. But being established professionals they command larger fees. At the other extreme you find the relatively untrained, mass production "Tax Experts". Their primary attraction is a low fee, although most clients end up paying considerably more than the advertised amount. H & R Block and other chains are usually somewhat in the middle.

If such a situation exists in your community, we suggest that you establish your fee level somewhere near those of the chains. As a newcomer you cannot demand the same fee as a long-established professional. On the other hand, to equate yourself with the untrained, so-called, 'tax expert', can only serve to downgrade your reputation and public confidence in your abilities.

Remember: by virtue of your specialized professional training you are entitled to professional fees, providing you render a professional service. And once you are established and become known as a capable tax specialist you will be able to raise your fees accordingly.

b. The amount of work and time spent in the preparation of a return. If a client comes to you with a complete summary of income and expenses, you will charge less—other things being equal—than if you have to reconstruct his financial history item by item. If you come to a client's home or business you are entitled to extra compensation for your traveling time and trouble. At the beginning you may be a little slow and may even make some time-consuming errors, it would be unwise and unfair to charge for time lost due to your lack of experience.

You will also find that it takes considerably longer to prepare a return for a first time client than for a client you have served before. Once you have accumulated certain basic information—familiarity with his business operations or other sources of income, special problems in connection with deductions and exemptions and so on—tax return preparation time will be drastically reduced. Since you expect the client to return year after year, you would be well advised to keep this in mind in setting your fee.

c. The value of your services. This factor provides you with your greatest opportunity for sizable fees. If the client is convinced that you effected a sizable tax savings for him, he will not object to your fee. Especially valuable in this respect is a tax-refund, and with care and patience you should be able to effect such savings and get refunds for many of your clients. With a little practice you will be able to tell, in most cases, which of your clients offer opportunities for worthwhile savings. The returns of lower income workers, for instance, with no outside income or business activities, are usually not complicated. Some of these taxpayers could just as well have prepared their own returns and you can't expect a big fee from them. On the other hand—business and professional people, investors, taxpayers with income from real estate rentals and sales, sales reps, and others who use their car on business—these frequently afford good chances for large deductions, and you should spare no time or effort to ferret out all possible savings.

Most preparers set a minimum fee for a relatively simple standard deduction return. For itemized returns requiring additional schedules or statements, or for more complicated and time-consuming work, they charge for each additional form or schedule. In such cases we suggest that you submit an itemized bill to explain your charges. Also, for corporation, partnership, and estate tax work, compensation is usually much higher than for individual income tax work, since more specialized knowledge is required.

d. The client's income. Another factor to consider when setting a fee is the size of the client's income. At first glance it may seem unfair to charge a higher fee just because a client has a higher income. The rationale for the additional charge, however, is that there are many instances where the tax practitioner is not fully compensated for the amount or difficulty of the work involved, simply because he realizes that the client can't afford it. Furthermore, the chances of an audit generally increase with the size of a client's income, which means that greater attention and care must be given to a higher income return.

In setting fees, both the consultant and the client should be aware that the taxpayer is paying for the consultant's professional skill and knowledge as well as for his time. A well-trained professional tax practitioner will certainly be able to prepare a tax return in such a way as to save the client considerably more than his fee entails. Obviously, the reason even 'tax-savvy' taxpayers hire a professional is because they realize that they cannot do the job as well as he can. The professional tax consultant should in turn be amply paid for his time and expertise.

With a little experience you should be able to judge how high a fee you can set, without making the client feel that you are overcharging. But until you are well established, it's wiser to undercharge rather than overcharge. Bear in mind that as a beginner you usually have less expense, less overhead—and less bargaining power with clients.

Fees for Extra Services

Every practitioner has a number of clients who repeatedly contact him during the year, with questions or with tax related problems, and so on. The extra time spent on one such client during the year, in addition to the time spent preparing his return adds up to a sizable amount, frequently overlooked when the tax return fee is established. Many practitioners keep a running log for such clients, entering every phone call, research study, letter or memorandum written, and so on. These items are tallied and an extra charge is made, or the return fee is appropriately increased to cover the additional time and work.

Often, you find that problems arise in the preparation of a client's return calling for specialized research. If the problem is of a general nature, many practitioners feel that the time required for the extra research should be considered as a general cost of keeping current. But, if the problem is a very specialized one, you have every right to expect payment for the additional time.

Also, if you spent time with—or for—your client in conducting year-end tax planning before the tax return season, you are obviously entitled to an additional fee.

Quite often, in connection with preparing a return, a client will ask you to perform some non-tax work for him as well. The tasks put in your lap may range from obtaining a refund for an unused plane ticket to helping straighten out a dispute with the electric company. In any case, if it involves an appreciable amount of time you should get paid for it.

Fees for Tax Examinations

Inevitably, some client returns will be selected for IRS and state tax examinations. This, of course, entails extra work on your part, either in representing the client at the audit or helping him prepare for it.

While a few tax practitioners do this type of work without extra charge, the great majority do charge extra for such services and we wholeheartedly concur.

If you absorb the additional time spent on audits you will eventually have to increase your fees to make up for the lost time. This, of course, means that all clients—audited or not—will have to pay extra. Moreover, a practitioner who represents a client at an audit, or spends time preparing for it without charging extra, conveys the impression that the audit is really due to his fault or negligence. This, in most cases, is not true, as we will explain later in the chapter dealing with the subject.

In truth, many seasoned tax consultants not only charge for representing clients at audits, but charge a premium fee, because audits call for special skills. Luckily, these are skills that you can easily acquire and put to very profitable use. Fees for preparing, or representing a client at an audit are generally charged on a time basis though a minority of practitioners charge a fixed minimum, with an additional charge if excessive time was required. In any event, if you succeeded in saving your client a substantial amount of money, a proportionately higher fee is justified.

Fee Schedules

Virtually all national or regional Tax Services, as well as many accountants and independent tax practitioners, use a fee schedule as a basis for setting fees. In many cases the fee schedule is intended to show the minimum fee only for a particular type of service, return, or schedule. These fees are often adjusted upward as the occasion demands.

A recent survey of tax consultants indicated that more than 90 percent have established minimum fees, 80 percent of these increase the minimum fee as the client's income increases, and almost all increase the fee for extra services rendered, such as specialized research. Another interesting point: about 30 percent of those surveyed said they increase their fees if their services result in tax savings to the client. The survey also revealed the fact that accountants charge more for tax work than for routine accounting work, because they feel that extra skill and knowledge is required for performing a proper tax service.

The survey also confirmed our finding that fees charged for the same service tended to increase in direct proportion to the number of years the tax practitioner has been in business.

Another interesting survey was undertaken by the National Society of Accountants (NSA) to determine average minimums and average fees charged by NSA members. The results were broken down into geographical area and population size according to the practitioner's locality.

As you can see from the results tabulated in the Appendix to this volume, fees do vary between regions and are somewhat greater in larger cities than in smaller communities.

You will also find a detailed, suggested fee schedule in the Appendix.

Chapter 11
COLLECTING YOUR FEES

You will find that the great majority of clients will pay your fee upon receiving their return or shortly thereafter. But, no doubt, you'll come across some individuals who are always late in paying their bills. The following pointers should help speed up your collections.

First and foremost, make it a practice to submit a formal bill to either accompany the tax return or very soon after. It is axiomatic to all collection efforts that the sooner the bill is received the sooner it is paid.

Unless you expect immediate payment, when the return is delivered or picked up by the client, we suggest that you include a self-addressed stamped envelope. Moreover, we feel that an envelope with a postage stamp is preferable to a business reply envelope. Somehow people find it harder to disregard an envelope bearing a valid stamp as opposed to an unstamped reply envelope. Indicate on your bill that you expect payment within 10 days (or 30 days, at the most) after receiving the bill.

As tactfully as you can, try to obtain payment when the return is delivered. For this reason, many practitioners do not mail returns to clients but have them pick up their returns in person—at which time the bill is presented. A good opportunity for payment is when the client must write out a check for additional taxes due or for estimated taxes. While the checkbook is yet in hand, it's easier to get him to write you a check as well. If he is short of funds, you could suggest a post-dated check, but tell him you would like to avoid the nuisance of having to send statements.

Some clients will tell you, "I'll pay the bill as soon as I receive my refund". This practice should be thoroughly discouraged. Like any other professional, you are entitled to compensation upon completion of your service, i.e. when the return is prepared and handed to the client. Once your fees become subject to a tax refund, any number of things can happen, and you may receive payment very late or not at all. By agreeing to wait for a refund you are, in effect, guaranteeing this refund, which you should avoid at all costs. The only exception to this is where both the bill and the refund are substantial, and you know that the client is simply unable to pay at the present time. Even then, we suggest that you and the client agree on a deadline (perhaps 90 days) by which you are paid regardless of the refund. Better yet, suggest that the bill be paid in installments, with the understanding that the entire balance be paid if the client receives his refund before all of the installments are due.

A growing number of accountants and tax preparers accept one or more of the major credit cards in payment of their services. The quicker collection will more than offset the small service charge.

Make sure to remind the client either by phone or by mail, no more than 30 days after the due date—if his bill is still unpaid.

If you serve the client on a year 'round basis, preparing his quarterly business tax returns or sales tax returns, you should expect payment on a quarterly basis—or even monthly—if the fee is large enough.

How to Bill

There are various schools of thought concerning how much itemization—if any—should be put on a bill. Current practice runs the gamut; from the casual sentence, For services rendered in preparing Federal and State individual tax returns, to an explicitly detailed bill listing every form, schedule, phone call, and other service rendered.

The most common practice, probably lies somewhere in between. Obviously, if you detail the bill, the client will better understand your charges. On the other hand, you don't want to appear too mercenary. But if you do prepare detailed bills, make sure to keep a copy in your files—as a reference in case your client questions you, and as a basis for billing in the future.

Latest figures show that about 65% of practitioners provide at least some itemization on their bills.

How to Handle Complaints About Your Fees

Of course, the best way to handle complaints is to avoid them. But short of charging so little that no client will ever object, you really have no way of forestalling all complaints. In the case of bills that are larger than usual, you can minimize complaints by drawing up an itemized account explaining the work you performed. This is most effective where special services were rendered to the client.

Another way is to explain to the client personally—either at billing time, or when the complaint arises—exactly how much work, research, time and skill went into preparing his return. Some will take out the client's file, go through all the papers, and point out exactly what was done.

Some tax consultants offer a cash discount (perhaps 5%) to all clients who pay their bills within a specified time, such as 10 or 30 days. Others feel that this method lacks professionalism. It may also help to add a reminder your client that your fee is fully tax deductible.

Be sure to take note of any comments the client makes about the fee on his record for future reference. If you know that a particular client is sticky about the fee (and assuming you want to keep him), you will be more careful next time in setting the fee, or at least in making sure you fully explain the charges when drawing up the bill.

Sometimes an entire return must be redone because the client omitted some important information. It is perfectly proper for you to charge for the additional time required, but make sure to note this fact on the bill and on your records; so you won't inadvertently charge an extra fee again next year.

As a general rule, whenever you present a bill that appears higher than the normal fee, take the initiative and explain the reasons for it. Do not put off your explanation until the client raises questions about the fee. Some clients may pay the bill without a word, but with a silent vow to never return again.

Some General Thoughts About Fees and Collections

A knowledgeable practitioner once remarked, "There is one sure way to recognize an inexperienced or unsuccessful professional, he's the one who usually tries to bypass or minimize the subject of fees. This apologetic attitude about fees can give the client the impression that you are afraid or ashamed to talk about them because they are out of line."

In other words, if a client questions your fee, don't go on the defensive. Point out that—though he may get cheaper service elsewhere—"you generally get what you pay for". Explain that you provide professional personalized, reliable service, that you are available year 'round, not just at tax time, and that the savings in taxes and aggravation may far outweigh the relatively minor difference in the fee.

Finally, you should consider dropping those clients who remain stubborn, even after you explain the reasons for your fee. Remember, those who pick a tax practitioner primarily on the basis of fees are not the ones who contribute much to your practice in the long run. Their loyalty is skin deep. They will probably drop you as soon as they find someone cheaper, and any clients they recommend will probably also expect an extra low fee from you.

Very often, a client will ask you in advance about your fee. If you quote your minimum fee—though you emphasize the fact that it is just a minimum—you will find it difficult to present a larger bill, even where you had more work than anticipated. In effect, you will generally find that your minimum fee becomes the maximum. But if you did quote a fee and then find that there is a lot more work involved than you had expected, it helps to give the client a breakdown of the additional time and charges. You could also, when quoting an estimate, advise the client that the fee will necessarily be larger if—in preparation of the return—you run into any unusual problems that require extra work or research.

In cases where you were successful in saving your client a substantial amount of money, we suggest you attach a memo to your bill explaining exactly what you did and approximately how much you saved. This will not only enhance your professional image and the client's respect for your abilities, but will also justify a larger fee.

Some Additional Billing and Collection Tips

The following suggestions should help you streamline your billing and collection procedures. If properly carried out, they should result in faster and easier collections, while maintaining the good will of your clients:

- Bill clients immediately upon completion of the work. If you wait they may forget the extent and efficiency of your services.
- Use round dollar amounts. Even if you base your fees on flat time rates don't send a bill for "$135.85".
- In writing out a detailed bill don't use technical abbreviations. Their meaning may be clear to you or to anyone in the field, but may be of no value to the client.
- Make sure to charge for long distance phone calls or for other out-of-pocket expenses (unless the amount is inconsequential in relation to the fee). These charges can add up very quickly.
- Make sure to follow up every unpaid bill after 30 or 60 days. Be courteous but persistent if payment is not forthcoming within a reasonable amount of time.
- Add a handwritten note to the bottom of the reminder statement, to give the payment request a sense of urgency.
- If the bill is large, suggest part payment to the client. But try to get a definite commitment as to when you can expect payment of the balance.
- Make sure to bill for non-tax work—especially if it was time consuming. Some clients are under the impression that tax practitioners don't charge for non-tax work.
- If you think the fee will be more than the client expects, prepare him before-

hand. Try to "sell," the fee by calling his attention to the amount of taxes saved, extent or difficulty of work performed, research required, etc.

- For efficiency in billing and follow-up notices, consider multiple-set, snap-apart statement forms. The second and third copies can serve as reminders.
- Instead of mailing a reminder notice, fax it. Faxes somehow convey a sense of urgency, without being unduly pushy. If the bill is seriously overdue, add a handwritten 'Please' or other note.

Chapter 12

OBTAINING ADDITIONAL INCOME FROM YOUR CLIENTS

While preparing income tax returns for individuals and businesses can be a very lucrative occupation, it is largely—but not entirely—seasonal. The figures vary, but many small to mid-sized practices handle about 70%-75% of their annual return volume during the January-April season, another 15%-20% during the May-October extension period, with the balance (mainly fiscal-year business returns) the rest of the year.

If this suits you, fine. But if you, like so many others, are looking for a year 'round business with steady year 'round earnings, you're in an excellent position to expand your services and revenues into a profitable, full-time, all-year source of income.

This chapter will alert you to a number of such opportunities. Bear in mind that these are not get-rich-quick schemes that'll put thousands of dollars into your pockets overnight. What we'll present here are practical, proven, ideas and service opportunities—both traditional and nontraditional—you can use to boost and grow your practice.

Some of these opportunities will come knocking on your door, e.g. clients, or prospective clients, will ask for them. Others will require your initiative in one of two ways: (1) Clients or potential clients looking for a service provider have to be made aware that you're capable and available, or (2) they have to be made aware of their need for your services.

It may take time and patience to develop a really lucrative practice.. But if you have the will and are prepared to put in the necessary effort and perseverance, your eventual success is almost assured.

Basically, once you go beyond preparing tax returns, the types of additional services you can offer fall into two categories: (A) tax (federal, state and local) related, and (B) nontax related.

We'll discuss the tax area first. This is probably the area with the quickest return on your efforts in time and effort.

Overview

Besides the preparation of a yearly tax return, many taxpayers find themselves in need of assistance with tax problems at various times throughout the year. A number of returns and information reports affecting many taxpayers—especially business people—are due during the year. In addition, decisions and transactions must be made, which may have a favorable or unfavorable effect on taxes. The average taxpayer is not familiar with the intricacies of tax law and is only too glad to dump the burden into the lap of a tax advisor. That advisor can certainly be you! Encourage your clients to consult with you about the tax effects of every important transaction: purchase or sale of stocks and bonds; purchase, sale, and improvement of real estate; large business expenditures, contributions, and so on.

Make it a point to assist, or advise your clients of your availability for counsel throughout the year, and not just during tax season. Needless to say, you must stay up-to-date on current tax law and related developments, and make sure your clients are well aware of this.

Tax Planning

Probably, the most profitable area in the tax field is tax planning. This is true, both in terms of financial remuneration—and more importantly—in terms of increasing your clientele, enhancing your professional image, and building a genuinely successful tax practice. Obviously, tax planning involves in-depth tax work that goes far beyond the routine service rendered by the average tax preparer.

Tax planning calls for a continuous, abiding interest in the client's financial and personal situation, combined with a thorough knowledge and understanding of tax law, and an awareness of likely changes. Tax planning requires imagination, a sympathetic attitude, and creativity.

Tax planning, more than anything else, results in client referrals—from associates, or friends of your clients who are in the same business, profession, or who have similar jobs. This, of course, enables you to get multiple mileage from your initial effort, because you can employ the same ideas or suggestions for some of the other clients as well.

As one knowledgeable practitioner put it: "A competent, dedicated tax consultant is in a position to become as indispensable to his clients as the family doctor once was to his patients, provided he maintains a close, personal, working relationship with his clients." This means that you should get to know them, both as clients and as individuals, and that you should be aware of developments in both their business or professional ,and in their personal lives.

Expanding Your Tax Services

Bear in mind that there are various levels of tax planning, requiring various levels of skills. Keeping abreast of the current tax literature, studying a few of the annual tax guides that usually spell out the more common tax saving ideas and steps, plus reading a good book or two on tax planning (make sure it's current) should equip you to service most clients.

Wealthier taxpayers—especially those with income (and losses) from varied sources—often require more sophisticated, customized, long-term planning with multi-year 'what-if' projections—at correspondingly higher fees, of course. There is tax-planning software available that makes this task much easier and quicker. If you have—or are taking steps to acquire—the knowledge and skills to serve this extremely lucrative market you can expect to earn top dollars.

If you employ assistants, insist that they cultivate a similar attitude, and also—in an unassuming way—make sure your clients sense this concern.

Some practitioners use the slower, post-tax season months to pull out client files, one by one, and examine them closely for any simple avenues for tax savings. Even if only one out of ten such examination results in a tax advantage to a client, your time will have been well spent.

Another opportune time for tax planning is in the last 30—60 days before the close of the year—either the calendar year, or in the case of fiscal year clients, the close of the fiscal year. Some tax consultants have a practice of sending out a routine form letter to all clients with an income above a certain level—perhaps $40,000 to $50,000. In keeping with the adage that "an ounce of prevention is worth a pound of cure", this reminder informs clients that tax saving moves made before the end of the year may far outweigh any tax savings you can effect for them later during tax return time. If you can specify areas of particular interest—either to that client, or areas that have witnessed important changes during the year—then your letter will have even greater impact.

Obviously, even if the client does not respond to your invitation, it still serves as a valuable—and perfectly ethical—reminder to call you when tax time comes along.

The preparation of out-of-state returns is an especially big business near military installations and in and around border areas where a good portion of the population commutes to work in another state. By the same token, tax consultants working in these areas will pick up quite a bit of business from residents of neighboring states who travel to work.

While the tax preparation fee normally covers both the federal and your home state (plus local, if any) return, preparing out-of-state income tax returns calls for an additional fee. In fact, most practitioner charge proportionally more for out-of-state returns.

Taxpayers who moved in during the year from another state or worked out of state are often ignorant of their filing responsibility and must be made aware of the requirements.

An individual who resided in several states during the year may need to file in each state.

It's a good idea to familiarize yourself beforehand with the forms and instructions of those states you anticipate filing for. The topic of obtaining out-of-state tax forms and instructions was discussed in Chapter 3.

Extra Income from Handling Tax Examinations and Disputes

As we point out in a later section, it is virtually impossible for a tax consultant with any size practice to avoid IRS audits and controversies. Although some feel that it is their responsibility to follow through without additional charge when a return prepared by them is questioned, the great majority of practitioners treat it as a separate task and charge accordingly. Most of those who do not charge for audits are forced to increase their return preparation fees to those returns that are audited or questioned.

Inasmuch as this type of work often requires thorough research and personal negotiations with IRS agents, the fees charged for such services, are generally 25% to 50% higher than comparable time spent in the preparation of a return. If handled properly, this work can turn into an important source of additional income for you.

Another source of extra income for tax professionals is in the area of state tax audits. With the almost universal application of state income taxes and with ever increasing state tax rates, the incidence of income tax audits by state tax departments is steadily increasing. There is also more cooperation between the IRS and state tax administrations whereby the audit load is divided between the Federal government and the various states. Under this program some tax returns that would have normally been examined by IRS are assigned to state tax auditors instead. Any deficiencies that turn up on the state tax return are then reported to the IRS, which in turn, revises that individual's Federal Income Tax Return to reflect the disallowed deductions, additional income, etc.—as determined by the state.

For this reason, clients should be cautioned to approach a state audit—though the dollar amounts are usually not too high—with the same care they would give a Federal examinations.

Here again, if you prepared the return originally, you will most likely be asked to defend it as well.

Social Security Counseling

Anyone engaged in tax practice for any length of time will at some point be asked to assist clients or their relatives in social security matters. During the half-century of its existence the old Social Security Act has mushroomed into a gigantic benefit program that intimately affects the lives of nearly all Americans. Moreover, about half of the population now pays more in social security taxes than in Federal income taxes with the percentage due to increase in the future.

With more individuals becoming aware of their immense stake in social security, the demand for help in this area is sure to increase dramatically in the next few years. Here is a sample list of the type of services you may be called upon to render:

- Assist with application for social security benefits.
- Help clients decide whether to file for reduced early retirement benefits.
- Help clients avoid or reduce loss of benefits because of excess earnings.
- Help clients obtain necessary documents for benefit applications.
- Pre-retirement planning.
- Counsel self-employed individuals on how to obtain maximum benefits.
- Design retirement plans to allow clients to earn income without jeopardizing their benefits.
- Help dependents and survivors collect maximum benefits.
- Help clients avoid problems in applying for disability benefits, etc.

This list, which could easily be extended, gives you a sample of the opportunities for service—and earnings—in this field.

For many types of assistance, such as helping with applications or filling out reports, very little in the way of specialized knowledge is required. However, those who desire to take on social security practice seriously should make an in-depth study of the pertinent Social Security laws and regulations—especially those affecting coverage, computation of benefits, and loss of benefits. The greatest demand for services, and greatest opportunities for substantial assistance is in the area of disability benefits. We can assure you that your time and effort will be well rewarded.

Operating a Year 'Round Payroll Tax (or Business Tax) Service

This is a major source of income for many tax practitioners. As you know, any business employing one or more workers is required to file numerous Federal and state wage reports and returns. There are Federal and state withholding, social security and unemployment insurance taxes, disability benefit taxes, compensation insurance, and a host of other related tax returns and reports that have to be filled out. Very often there are Federal excise tax returns, state or local excise and sales tax returns, various census reports—plus other returns, statements and reports too numerous to mention—that every business, large or small, is required to submit.

The alert practitioner—able to secure a few such business-tax accounts on a monthly or yearly basis—can net a comfortable, steady, year 'round income. The fee for this service, which usually includes preparation of the proprietor's personal income tax returns, depends on the size of the business and the number of employees involved. It may range from $200-$400 per year for a very small business, to $50-$200 or more per month for a larger business—employing 10 and 25 people. You can also anticipate being asked by many of your clients' employees to prepare their income tax returns.

If you are able to maintain a part-time bookkeeping service in connection with your tax service, so much the better. Many small businesses do not have enough work to hire a full-time or even part-time bookkeeper and would be glad to let you handle the work. Remember, that when you provide a bookkeeping and tax service, you are in a good position to obtain your clients' recommendations for similar work at other firms.

Many tax specialists first get involved in bookkeeping through the back door, so to speak, but then discover it to be a lucrative area that can yield a nice, steady, year-round income. What often happens is that a few small business clients come to you to have their tax returns prepared. While questioning them, you find that they have scanty, or no records, which means—either they, or you—have to go through the lengthy process of reconstructing the figures in order to get the data you need for the return. Since every business is required by law to keep accurate records, it is usually an easy task to persuade these individuals to (a) engage someone to do the bookkeeping for them, and (b) let you be the one to handle that task. Small entrepreneurs are characteristically impatient to get on with their business and hate paper-work; so that should help too.

Your fee for a complete bookkeeping, payroll, and tax service will usually start at a minimum about $100 a month and may climb much higher, depending on the amount of work, and complications involved.

To operate such a service, all you have to do is set up a simple manual or computerized record-keeping system designed to provide the figures and other data required by the particular client. Depending on the client's needs, and amount of time he has available; either he or you will enter the data periodically. The amount of detailed work you have to do and the size of the establishment will determine whether your work is to be done on a weekly, monthly, or even quarterly basis. As a rule, most bookkeeping accounts are handled monthly, but if you have to do all the entries, or even make out the payroll and accounts payable checks, and especially with manual systems, then your services may be required weekly.

Some public bookkeepers keep their clients' books in their own office, and obtain information from their clients through daily or weekly report sheets they pick up or have mailed to them.

In trying to sell a client on your bookkeeping service, emphasize the fact that you will then be obviously in a much better position to prepare an accurate, and tax-saving, income tax return—because his books are in order, and you are entirely familiar with his operation. And furthermore, it will be much easier to substantiate the figures on the tax return, in case of a tax audit.

Here is a suggestion: When preparing the tax return for a new client, who may be a candidate for either a payroll tax, or a bookkeeping and tax service, take extra pains to do an outstanding job. Make sure not only that the return is technically correct and accurate, but also convince yourself—and your client—that every possible tax-saving opportunity was indeed explored. This will give him an indication of the personalized, dedicated, and professional service he can expect as a client of yours throughout the year.

A good way to get new clients is to compile a list, from your classified telephone directory, of a few hundred small business establishments, in or near your community, and send them a printed card or letter, offering your services. Your advertising should state that you would be glad to visit anyone interested and discuss your service without obligation. Enclose a business reply card for this purpose, then follow up with a second and third mailing a few weeks apart. The best time to get new accounts is from December to February, so concentrate your efforts in these months.

Particularly good prospects for either a Business-Tax or a Bookkeeping and Tax service are new businesses who have not as yet hired anyone to fill these needs. You can learn about new business establishments through your local paper and just by keeping your eyes and ears open. Another way to obtain the names and addresses of new firms is by visiting your County or Town Clerk's office periodically. All new firms using a trade name must register there. (In most states this requirement also applies to new partnerships, regardless of whether or not a trade name is used).

Contact such new businesses by mail, or preferably, in person a short time after their opening and again, a week or two before the next quarterly payroll tax return is due.

Since most new business proprietors are reluctant to invest too much in what they deem "unproductive" expenses for "paperwork", some practitioners offer a special low introductory rate for the first year. Remember, if the business grows, as is often the case, your workload and fees will grow with it. Always bear in mind that a few small clients can quickly lead to some exceedingly lucrative accounts.

Non-Tax Related Income Opportunities

Individuals involved in work of a public nature are often asked to assist in preparing various official and semi-official papers and documents. And so, the range of services you will be called upon to perform is virtually limitless.

Just to give you some idea of this, there are many older individuals who need help with such simple matters as filing claims for Medicare reimbursement. There are low income individuals who need assistance in obtaining welfare payments, food stamps, Medicaid, and various government subsidies and loans. And then there are insurance claims, accident reports, and local real estate tax matters such as Senior Citizen exemptions. The list goes on and on.

While the fees for these services may not be very high, they can add up to a comfortable and dependable year 'round income; and many of those you help will turn into loyal tax clients and boosters as well.

We also suggest that you obtain a commission as a Notary Public. While notary fees alone rarely amount to a major source of income; many reports, statements, and applications, do require notarization—so being a Notary will help you secure more business.

In the last few years the Federal and state governments have become involved in numerous additional programs affecting much of the population—especially lower income individuals and families. This trend is going to accelerate as taxpayers demand and receive more assistance and services from the Federal, state and local governments, and the resulting paperwork is going to be staggering. This means that there will be a greater demand for those ready, willing, and able to help the average citizen cope with the increase in forms, applications, reports and statements to fill out. As you gain expertise with the various programs and their accompanying forms, you can expect the demand

Chapter 13
ETHICAL RESPONSIBILITIES OF THE TAX CONSULTANT

Aside from the legal requirements and responsibilities just discussed, the reputable tax practitioner should also be aware of the moral and ethical responsibility involved in handling a client's tax affairs. One common concern is whether—and to what extent—you must or should go to ascertain whether the return you are preparing for your client is a true and correct one.

The following rules and guidelines were issued by The American Institute of Certified Public Accountants (AICPA), the national professional organization of CPAs. While non-members are not bound by these rules, you may still wish to use them as a guide in your own practice.

Certain Procedural Aspects of Preparing Returns

.01 This statement considers the responsibility of the CPA to examine or verify certain supporting data or to consider information related to another client when preparing a client's tax return.

.02 In preparing or signing a return, the CPA may in good faith rely without verification upon information furnished by the client or by third parties. However, the CPA should not ignore the implications of information furnished and should make reasonable inquiries if the information furnished appears to be incorrect, incomplete, or inconsistent either on its face or on the basis of other facts known to the CPA. In this connection, the CPA should refer to the client's returns for prior years whenever feasible.

.03 Where the Internal Revenue Code or income tax regulations impose a condition to deductibility or other tax treatment of an item (such as taxpayer maintenance of books and records or substantiating documentation to support the reported deduction or tax treatment), the CPA should make appropriate inquiries to determine to his or her satisfaction whether such condition has been met.

.05 The preparer's declaration on the income tax return states that the information contained therein is true, correct, and complete to the best of the preparer's knowledge and belief "based on all information of which preparer has any knowledge."

This reference should be understood to relate to information furnished by the client or by third parties to the CPA in connection with the preparation of the return.

.06 The preparer's declaration does not require the CPA to examine or verify supporting data. However, a distinction should be made between (1) the need to either determine by inquiry that a specifically required condition (such as maintaining books and records or substantiating documentation) has been satisfied, or to obtain information when the material furnished appears to be incorrect or incomplete, and (2) the need for the CPA to examine underlying information. In fulfilling his or her obligation to exercise due diligence in preparing a return, the CPA ordinarily may rely on information furnished by the client unless it appears to be incorrect, incomplete, or inconsistent. Although the CPA has certain responsibilities in exercising due diligence in preparing a return, the client has ultimate responsibility for the contents of the return. Thus, where the client presents unsupported data in the form of lists of tax information, such as dividends and interest received, charitable contributions, and medical expenses, such information may be used in the preparation of a tax return without verification unless it appears to be incorrect, incomplete, or inconsistent either on its face or on the basis of other facts known to the CPA.

.07 Even though there is no requirement to examine underlying documentation, the CPA should encourage the client to provide supporting data where appropriate. For example, the CPA should encourage the client to submit underlying documents for use in tax return preparation to permit full consideration of income and deductions arising from security transactions and from pass-through entities such as estates, trusts, partnerships, and S corporations. This should reduce the possibility of misunderstanding, inadvertent errors, and administrative problems in the examination of returns by the Internal Revenue Service.

Another problem arises in the following situation:

Suppose you gave a client advice on a tax matter such as a proposed transaction or form of business organization. Then, a change in the law, new Revenue Ruling, or court decision affects your previous advice or even makes it inapplicable. Must, or should, you notify the client of these developments?

There is no question that you will enhance your standing in his eyes if you take the time and trouble to advise him of the change. But, whether you actually have a moral or ethical responsibility to do so is open to question. Here again, we draw on the guidelines provided by the AICPA to its members. We realize that these guidelines may not be entirely appropriate to your situation but they should surely help you to define your own course of action.

Form and Content of Advice to Clients

.01 This statement discusses certain aspects of providing tax advice to a client and considers the circumstances in which the CPA has a responsibility to communicate with the client when subsequent developments affect advice previously provided. The statement does not, however, cover the CPA's responsibilities when it is expected that the advice rendered is likely to be relied upon by parties other than the CPA's client.'

.02 In providing tax advice to a client, the CPA should use judgment to ensure that the advice given reflects professional competence and appropriately serves the client's needs. The CPA is not required to follow a standard format or guidelines in communicating written or oral advice to a client.

.03 In advising or consulting with a client on tax matters, the CPA should assume that the advice will affect the manner in which the matters or transactions considered ultimately will be reported on the client's tax returns. Thus, for all tax advice the CPA gives to a client, the CPA should follow the standards in SRTP No. 1 relating to tax return positions.

.04 The CPA may choose to communicate with a client when subsequent developments affect advice previously provided with respect to significant matters. However, the CPA cannot be expected to have assumed responsibility for initiating such communication except while assisting a client in implementing procedures or plans associated with the advice provided or when the CPA undertakes this obligation by specific agreement with the client.

.05 Tax advice is recognized as a valuable service provided by CPAs. The form of advice may be oral or written and the subject matter may range from routine to complex. Because the range of advice is so extensive and because advice should meet special needs of a client, neither standard format nor guideline for communicating advice to the client can be established to cover all situations.

.06 Although oral advice may serve a client's needs appropriately in routine matters or in well-defined areas, written communications are recommended in important, unusual, or complicated transactions. In the judgment of the CPA, oral advice may be followed by a written confirmation to the client.

.07 In deciding on the form of advice provided to a client, the CPA should exercise professional judgment and should consider such factors as the following:

a. The importance of the transaction and amounts involved
b. The specific or general nature of the client's inquiry
c. The time available for development and submission of the advice
d. The technical complications presented
e. The existence of authorities and precedents
f. The tax sophistication of the client and the client's staff
g. The need to seek legal advice

.08 The CPA may assist a client in implementing procedures or plans associated with the advice offered. During this active participation, the CPA continues to advise and should review and revise such advice as warranted by new developments and factors affecting the transaction.

.09 Sometimes the CPA is requested to provide tax advice but does not assist in implementing the plans adopted. While developments such as legislative or administrative changes or further judicial interpretations may affect the advice previously provided, the CPA cannot be expected to communicate later developments that affect such advice unless the CPA undertakes this obligation by specific agreement with the client. Thus, the communication of significant developments affecting previous advice should be considered an additional service rather than an implied obligation in normal CPA-client relationship.

.10 The client should be informed that advice reflects professional judgment based on an existing situation and that subsequent developments could affect previous professional advice. CPAs should use precautionary language to the effect that their advice is based on facts as stated and authorities that are subject to change.

What happens when you feel that existing rules or regulations do not accurately interpret the law? In addition to the legal implications already discussed, there are ethical considerations as to how far you may go in disregarding official IRS pronouncements.

We will once more rely on the AICPA for guidance in this area.

Tax return position

.01 This statement sets forth the standards a CPA should follow in recommending tax return positions and in preparing or signing tax returns including claims for refunds. For this purpose, a "tax return position" is (1) a position reflected on the tax return as to which the client has been specifically advised by the CPA or (2) a position as to which the CPA has knowledge of all material facts and, on the basis of those facts, has concluded that the position is appropriate.

.02 With respect to tax return positions, a CPA should comply with the following standards:

a. A CPA should not recommend to a client that a position be taken with respect to the tax treatment of any item on a return unless the CPA has a good faith belief that the position has a realistic possibility of being sustained administratively or judicially on its merits if challenged.

b. A CPA should not prepare or sign a return as an income tax return preparer if the CPA knows that the return takes a position that the CPA could not recommend under the standard expressed in paragraph .02a.

c. Notwithstanding paragraphs .02a and .02b, a CPA may recommend a position that the CPA concludes is not frivolous so long as the position is adequately disclosed on the return or claim for refund.

d. In recommending certain tax return positions and in signing a return on which a tax return position is taken, a CPA should, where relevant, advise the client as to the potential penalty consequences of the recommended tax return position and the opportunity, if any, to avoid such penalties through disclosure.

.03 The CPA should not recommend a tax return position that-

a. Exploits the Internal Revenue Service (IRS) audit selection process; or

b. Serves as a mere "arguing" position advanced solely to obtain leverage in the bargaining process of settlement negotiation with the Internal Revenue Service.

.04 A CPA has both the right and responsibility to be an advocate for the client with respect to any positions satisfying the aforementioned standards.

.05 Our self-assessment tax system can only function effectively if taxpayers report their income on a tax return that is true, correct, and complete. A tax return is primarily a taxpayer's representation of facts, and the taxpayer has the final responsibility for positions taken on the return.

.06 CPAs have a duty to the tax system as well as to their clients. However, it is well-established that the taxpayer has no obligation to pay more taxes than are legally owed, and the CPA has a duty to the client to assist in achieving that result. The aforementioned standards will guide the CPA in meeting responsibilities to the tax system and to clients.

.07 The standards suggested herein require that a CPA in good faith believe that the position is warranted in existing law or can be supported by a good faith argument for an extension, modification, or reversal of existing law. For example, the CPA may reach such a conclusion on the basis of well-reasoned articles, treatises, IRS General Counsel Memoranda, a General Explanation of a Revenue Act prepared by the staff of the Joint Committee on Taxation and Internal Revenue Service written determinations (for example, private letter rulings), whether or not such sources are treated as "authority" under section 6661. A position would meet these standards even though, for example, it is later abandoned due to practical or procedural aspects of an IRS administrative hearing or in the litigation process.

.08 Where the CPA has a good faith belief that **more than one** position meets the standards suggested herein, the CPA's advice concerning alternative acceptable positions may include a discussion of the likelihood that each such position might or might not cause the client's tax return to be examined and whether the position would be challenged in an examination.

.09 In some cases, a CPA may conclude that a position is not warranted under the standard set forth in the preceding paragraph, .02a. A client may, however, still wish to take such a tax return position. Under such circumstances, the client should have the opportunity to make such an assertion, and the CPA should be able to prepare and sign the return provided the position is adequately disclosed on the return or claim for refund and the position is not frivolous. A "frivolous" position is one which is knowingly advanced in bad faith and is patently improper.

.10 The CPA's determination of whether information is adequately disclosed by the client is based on the facts and circumstances of the particular case. No detailed rules have been formulated, for purposes of this statement, to prescribe the manner in which information should be disclosed.

.11 Where the particular facts and circumstances lead the CPA to believe that a taxpayer penalty might be asserted, the CPA should so advise the client and should discuss with the client issues related to disclosure on the tax return. Although disclosure is not required if the position meets the standard in paragraph .02a, the CPA may nevertheless recommend that a client disclose a position. Disclosure should be considered when the CPA believes it would mitigate the likelihood of claims of taxpayer penalties under the Internal Revenue Code or would avoid the possible application of the six-year statutory period for assessment under section 6501(e). Although the CPA should advise the client with respect to disclosure, it is the client's responsibility to decide whether and how to disclose.

Finally, every practitioner encounters situations where he discovers errors or omissions made by another practitioner. Is he morally or ethically bound to bring the matter to the clients' attention ?

Suppose he does, and the client refuses to correct the error, should or must the practitioner alert the IRS or state (or local) tax authorities?

Here's what the AICPA has to say on the subject. And we strongly endorse their position, whether you are an AICPA member or not.

Knowledge of Error: Return Preparation

.01 This statement considers the responsibility of a CPA who becomes aware of an error in a client's previously filed tax return or of the client's failure to file a required tax return. The term "error" includes a position taken on a prior year's return that no longer meets these standards due to legislation, judicial decisions, or administrative pronouncements having retroactive effect. However, an error does not include an item that has an insignificant effect on the client's tax liability.

.02 This statement applies whether or not the CPA prepared or signed the return that contains the error.

.03 The CPA should inform the client promptly upon becoming aware of an error in a previously filed return or upon becoming aware of a client's failure to file a required return. The CPA should recommend the measures to be taken. Such recommendation may be given orally. The CPA is not obligated to inform the Internal Revenue Service, and the CPA may not do so without the client's permission, except where required by law.

.04 If the CPA is requested to prepare the current year's return and the client has not taken appropriate action to correct an error in a prior year's return, the CPA should consider whether to withdraw from preparing the return and whether to continue a professional relationship with the client. If the CPA does prepare such current year's return, the CPA should take reasonable steps to ensure that the error is not repeated.

.05 While performing services for a client, a CPA may become aware of an error in a previously filed return or may become aware that the client failed to file a required return. The CPA should advise the client of the error (as required by Treasury Department Circular 230) and the measures to be taken. It is the client's responsibility to decide whether to correct the error. In appropriate cases, particularly where it appears that the Internal Revenue Service might assert the charge of fraud or other criminal misconduct, the client should be advised to consult legal counsel before taking any action. In the event that the client does not correct an error, or agree to take the necessary steps to change from an erroneous method of accounting, the CPA should consider whether to continue a professional relationship with the client.

.06 If the CPA decides to continue a professional relationship with the client and is requested to prepare a tax return for a year subsequent to that in which the error occurred, then the CPA should take reasonable steps to ensure that the error is not repeated. If a CPA learns the client is using an erroneous method of accounting, when it is past the due date to request IRS permission to change to a method meeting the standards of SRTP No. 1, the CPA may sign a return for the current year, providing the return includes appropriate disclosure of the use of the erroneous method.

.07 Whether an error has no more than an insignificant effect on the client's tax liability is left to the judgment of the individual CPA based on all the facts and circumstances known to the CPA. In judging whether an erroneous method of accounting has more than an insignificant effect, the CPA should consider the method's cumulative effect and its effect on the current year's return.

Knowledge of Error: Administrative Proceedings

.01 This statement considers the responsibility of a CPA who becomes aware of an error in a return that is the subject of an administrative proceeding, such as an examination by the IRS or an appeals conference relating to a return or a claim for refund. As used herein, the term "error" includes any position, omission, or method of accounting, which, at the time the return is filed, fails to meet the standards set out in SRTP No. 1. The term "error" also includes a position taken on a prior year's return that no longer meets these standards due to legislation, judicial decisions, or administrative pronouncements having retroactive effect. However, an error does not include an item that has an insignificant effect on the client's tax liability.

.02 This statement applies whether or not the CPA prepared or signed the return that contains the error; it does not apply where a CPA has been engaged by legal counsel to provide assistance in a matter relating to the counsel's client.

.03 When the CPA is representing a client in an administrative proceeding with respect to a return which contains an error of which the CPA is aware, the CPA should inform the client promptly upon becoming aware of the error. The CPA should recommend the measures to be taken. Such recommendation may be given orally. The CPA is neither obligated to inform the Internal Revenue Service nor may the CPA do so without the client's permission, except where required by law.

.04 The CPA should request the client's agreement to disclose the error to the Internal Revenue Service. Lacking such agreement, the CPA should consider whether to withdraw from representing the client in the administrative proceeding and whether to continue a professional relationship with the client.

.05 When the CPA is engaged to represent the client before the Internal Revenue Service in an administrative proceeding with respect to a return containing an error of which the CPA is aware, the CPA should advise the client to disclose the error to the Internal Revenue Service. It is the client's responsibility to decide whether to disclose the error. In appropriate cases, particularly where it appears that the Internal Revenue Service might assert the charge of fraud or other criminal misconduct, the client should be advised to consult legal counsel before taking any action. If the client refuses to disclose or permit disclosure of an error, the CPA should consider whether to withdraw from representing the client in the administrative proceeding and whether to continue a professional relationship with the client.'

.06 Once disclosure is agreed upon, it should not be delayed to such a degree that the client or CPA might be considered to have failed to act in good faith or to have, in effect, provided misleading information. In any event, disclosure should be made before the conclusion of the administrative proceeding.

.07 Whether an error has an insignificant effect on the client's tax liability should be left to the judgment of the individual CPA based on all the facts and circumstances known to the CPA. In judging whether an erroneous method of accounting has more than an insignificant effect, the CPA should consider the method's cumulative effect and its effect on the return which is the subject of the administrative proceeding.

Chapter 14
TAX AUDITS AND HOW TO HANDLE THEM

How and Why a Return is Selected for Audit

Every tax return filed is checked for mathematical accuracy and completeness. If any mathematical errors are found, or if some vital information or required attachments are missing, the taxpayer will receive a notice, either correcting the error or asking for the missing data. Next, the return is screened by the IRS computer for its "audit potential". The return's audit potential is rated on the basis of a mathematical formula called the Discriminant Function System (DIF), under which various weights are assigned to different entries on the return. The exact characteristics that will flag a return are, of course, kept secret. But generally speaking, any income tax return that shows larger than usual personal deductions for the amount of income reported, especially in such categories as contributions, medical expenses, etc., as well as certain types of business expenses, particularly large travel and entertainment expenses, is a likely candidate for an audit. Nevertheless, if the examiner—on the basis of his experience and judgment—feels that the return appears to be in order, it will be returned to the file; if not, the return is forwarded to the IRS Audit Branch nearest the taxpayer's residence.

In addition to the computer screening test, IRS also annually selects a certain percentage of returns at random for further examination. The exact percentage varies from year to year. In the last year for which figures are available the range was from less than one-half percent for returns in the under $20,000 income group, to almost 9% in the case of returns reporting more than $100,000 in income. The percentage is also higher for returns showing itemized deductions than for standard deduction returns and for returns with Schedule C business income.

The IRS also conducts ongoing taxpayer compliance programs in which it annually examines almost every taxpayer in selected groups, businesses or professions, in a particular area. Often targeted are taxpayers, such as doctors, cab drivers, waiters and others who receive a portion of their income in cash.

There are other reasons too, why a return may be selected for closer examination—usually as a result of manual screening by IRS examiners. For instance, a taxpayer reporting poverty level income, but residing in a "posh" neighborhood is a likely target. Or, the individual may have had his return done by a taxpreparer who is suspected, or being investigated, by the IRS for unscrupulous or fraudulent practices. Other factors that may trigger an audit are large refund claims, the taxpayer's prior history, discrepancies between the return and information documents filed with IRS (W-2s, 1099s, etc.), or information received from informers In recent years IRS frequently conducted so-called "financial status audits" and "economic reality checks" to flush out unreported income. Agents would examine in detail the taxpayer's assets, living style, and estimated expenditures to determine whether they match the income reported on the return. The 1998 tax law restricts these techniques to situations where the IRS has other indications of unreported income.

According to latest government figures; of the over 96 million returns filed in a recent year, 1.7 million, or 1.8%, were selected for audit. Of these, roughly two-thirds were picked because they matched the computer profile for returns with high audit potential; the balance were selected manually or in connection with one of the various enforcement programs.

But no matter what the percentage odds are, you can expect that some of your clients will be audited and the great majority will turn to you for assistance. If you are new or relatively inexperienced in this area you're likely to be somewhat apprehensive. The following discussion, advice and hints should alleviate your apprehension, and whether you are a novice or a seasoned practitioner, these suggestions will help you face the mighty IRS with confidence and substantially improved prospects.

Reassure Your Client

Your first task is to reassure the client that the selection of his return does not necessarily mean that anything is wrong, that it is due to any negligence or error on his part or yours, or that the IRS suspects him of cheating. Explain the selection process to him, emphasizing that the return may have been picked at random, or that he may be in a particular class of taxpayers being checked. Make it clear to him that an individual (especially with more than average income) whose return has never been examined may very well have been overpaying his tax. This observation is particularly appropriate when you are dealing with a new client.

To remove some of the dread a tax examination brings on, it may help if you acquaint your client with the official "IRS Statement of Principles of Tax Administration", which are supposed to guide all IRS employees in their dealings with the public:

"The function of the Internal Revenue Service is to administer the Internal Revenue Code. Tax policy for raising revenue is determined by Congress.

"With this in mind, it is the duty of the Service to carry out that policy by correctly applying the laws enacted by Congress; to determine the reasonable meaning of various Code provisions in light of the Congressional purpose in enacting them; and to perform this work in a fair and impartial manner, with neither a government nor a taxpayer point of view.

"At the heart of administration is interpretation of the Code. It is the responsibility of each person in the Service, charged with the duty of interpreting the law, to try to find the true meaning of the statutory provision and not to adopt a strained construction in the belief that he is "protecting the revenue".

"The Service also has the responsibility of applying and administering the law in a reasonable, practical manner. Issues should only be raised by examining officers when they have merit, never arbitrarily or for trading purposes.

"Administration should be both reasonable and vigorous. It should be conducted with as little delay as possible and with great courtesy and consideration. It should never try to overreach and should be reasonable within the bounds of law and sound administration."

Type of Examination

The IRS has two examination procedures, and the type of procedure selected depends largely on the type of return and size of income. Individual returns, especially those in the lower income brackets, are normally selected for office examination (formerly called an "office audit"). This means that the taxpayer (or his representative) is requested to appear at the nearest IRS office at the designated time and date, together with the required records and further substantiation. Sometimes, an office examination is handled by correspondence and the taxpayer is requested to send the required data by mail. This method is usually employed where the desired data or substantiation is not too extensive.

A field examination (previously called a "field audit"), where the IRS agent comes to the taxpayer's business, office, or home is usually confined to business tax returns or high income individual taxpayers. Occasionally, a field examination is set up where the IRS suspects that a taxpayer with a small reported income lives in a lavish style, way out of proportion to his reported earnings. The examiner then—besides checking taxpayer's books—will take a close look at the type of home he lives in, his manner, and standard of living, etc.

In the usual case, though, where you are dealing with an individual, non-business return, your client will be involved in an office examination. Assuming that the taxpayer was asked to personally appear at the IRS office, several possibilities arise: (1) The client can go by himself; (2) You can go with the client; or (3) You can go alone as the client's representative (subject to the rules of practice previously discussed). The route to choose depends on the complexity of the return, the number of issues involved, and equally important—the nature and temperament of the client.

It it's a relatively simple return, and all or mostly all substantiation is available, there is really no need for you to go to the examination, unless the client is temperamentally unfit to go by himself. There are individuals who, for some reason, become extremely agitated, frightened, or antagonistic when confronted with IRS agents. Others get themselves into hot water by talking too much.

If it is decided that you will represent the client, the next question is whether or not to take him along. Here again the previously mentioned factors should be taken into account.

If the client is temperamentally suitable and willing to go, we suggest that you take him along. For one, it may hasten the examination because there will be a number of questions and issues raised that only he can answer. There is also a psychological advantage to his being there. Frequently an agent may doubt a client's statements or representations. He may even suspect some of the records as being "doctored". Nevertheless, he will usually hesitate to call the client a "liar" to his face, while he will have no such reservations in the latter's absence. Likewise, if the client is obviously unsophisticated in these matters, or convincingly appears as honest and trustworthy, it may help to put some of the agent's doubts to rest.

Find Out What the IRS Wants

Your first step in preparing for the examination is to find out exactly what the IRS is after. In most cases, the letter advising the taxpayer of the examination gives that information. The letter may request substantiation of contributions, medical expenses, rental expenses, and so on.

But your quest should go beyond that. Oftentimes, one or two items on the return triggered the examination, and if these can be explained and substantiated to the satisfaction of the examiner, you may have smooth sailing as far as the rest of the return goes.

You can sometimes, but not always, get this further information by simply calling the examiner in charge and discussing the matter with him. In fact, many experienced practitioners routinely call the agent, whenever a client is audited, to discuss the matter with him. Ostensibly, many call to set up an appointment or to change the appointment, but their real purpose is to determine what the agent is really after. If the agent is unknown to you, and you plan to either accompany the client or represent him at the examination, (more about that later), it will not hurt if you speak to him beforehand and try to establish some kind of personal rapport with him.

Generally, the IRS agent will not discuss a client's return with you, unless he has on file a Power of Attorney (on Form 2848 or a written statement) from the client authorizing you to act on his behalf.

If you are lucky, you can determine during your preliminary conversation whether he is the "nit-picking" type who will insist that you show proof for every penny claimed; or whether he is the more broadminded type who will concentrate on the big issues, and not spend much time fretting about petty items. Fortunately, the latter type are not in the ascendancy. At the same time, you can try to impress the agent with your knowledge, expertise, and professionalism. From the beginning give him to understand that you have every intention—and are capable—of backing up the return, and that you have full confidence in your client's position.

Preparing for the Examination

Now down to the nitty-gritty. Assuming that you have a fair idea of what the IRS is after, it is up to you to help, coax, perhaps even needle, your client into providing the necessary information. After all, the return is his, not yours, and it is imperative that your client understand that the responsibility squarely falls on his shoulders, not on yours. You can help and assist him in every possible way, but it is his task to assemble the necessary bills, receipts, checks and other supportive items.

Some clients take the attitude: "My tax consultant prepared the return, let him worry about it." Needless to say, this is an unhealthy, counterproductive attitude and you should seek to eradicate it—not when the return is examined, but when it is initially prepared. By the time the return is examined, it is usually one to several years after the return was filed.

We suggest that you assemble the return under examination as well as all related papers, copies of previously audited returns, etc., into a separate file and keep them all together. Never mix audit papers with your current year's worksheets or returns.

What do you do if the client is unable to produce any kind of substantiation for a large expenditure? Here, a full explanation may be helpful. In one case, for instance, a substantial deduction for medicines and drugs was originally reduced to $50 because the taxpayer was unable to show receipts or checks. The tax preparer then prepared a letter explaining that one of the taxpayer's children (identified by name) was required to take regular daily anti-allergy medication, at a cost of so and so-much per week. He also listed the name of the drug, the dosage and the name of the physician. Other drug items were similarly explained with the result that the full deduction was reinstated without further question.

As a rule, the more "meat" you can furnish in the way of names, dates, exact amounts, reasons, and explanations, etc., the better your chances are of seeing the questioned deductions go through. However, while preparing for the examination, if you come across weak spots in the return—items you feel you will have a hard time substantiating or defending—by all means, tell your client right away to avoid disappointment later on. It is better for you, and the client, if he expects the worst and then is surprised to find that it turned out better than expected, than to be confident at first only to experience a rude awakening later.

Incidentally, if the return being examined was prepared by someone else, or the client himself, you have a good chance to convert that taxpayer into a regular client. Don't knock the previous preparer, but show how you would have maintained a stronger position vis-a-vis the IRS, if you had prepared the return—given the precautions you normally take.

Coaching the Client for the Examination

Let's assume the client decides to attend the audit by himself, rather than have you accompany or represent him. A little coaching session with you beforehand will greatly enhance his chances.

First, and foremost, impress the client with the necessity of being cooperative, not belligerent. Human nature being what it is, the taxpayer's obstinacy will tend to make the agent more suspicious and more determined to dig into every item under examination. It may even invite him to raise new issues. At the same time, it is essential that you advise your client not to volunteer any information. Let him answer all questions, furnish all requested data and records, but no more! Many a taxpayer comes to grief because, in his eagerness to be extra helpful or impress the agent with his honesty, he unwittingly provides him with damaging information or clues. As mentioned previously, if—in your judgment—the client is either too belligerent or too voluble, better dissuade him from going to the examination; or at least make sure that you accompany him and try to keep him under control. If the client is intelligent enough and knowledgeable, you can also brief him on some of the points discussed in the next section.

How to Deal with the IRS Agent

In dealing with an agent who is examining a client's return, bear in mind that his job is to protect the interests of the Treasury by carefully examining the return to determine whether it shows the full legal tax liability. As a result, the average agent develops a rather suspicious nature. However, it would be wrong to treat every examiner from the start as an adversary. In fact, the IRS in recent years has taken pains to keep the previously antagonistic client-agent atmosphere out of examinations and conferences with taxpayers.

IRS agents, like all human beings, vary in nature, temperament and attitude. Though, by official policy, they are supposed to determine the tax due with reasonableness and fairness, rather than squeeze out every possible dollar, some are more zealous or technical than others. But in general, if you give the agent the feeling that you are dealing with him in good faith, there's a good chance that he will reciprocate and treat you and your client likewise.

It is therefore up to you to avoid turning the agent into an opponent. Be friendly, without indulging in backslapping. Treat him in a businesslike manner, show that you respect him, that you realize he has a job to do. On the other hand, your attitude should clearly read that it is your job to protect the taxpayer and that you are not afraid to do so. In other words, establish at the outset an atmosphere of mutual respect. This is especially important if you live in a rural area or small town, since you are likely to come into contact with the same few agents over and over. Therefore, it is wise to establish this good rapport as early as possible. Bear in mind too, that agents, among themselves, indulge in a certain amount of "shop-talk"—discussing cases, taxpayers, and tax practitioners. This means there is a good possibility that an agent you haven't met as yet has already heard about you—especially if the job you do is out of the ordinary.

Having attended to the preliminaries, the agent will go over the items in question one by one. If the return is "tight" and you have sufficient documentary evidence to support your position, then you have little to worry about. However, if—as in most cases—some or many items are in the "gray" area, you will need tact, diplomacy and a certain amount of psychology to come out ahead.

Although IRS agents are specifically enjoined from engaging in "horsetrading", it frequently takes a considerable amount of bargaining—involving give-and-take—to arrive at a satisfactory settlement, particularly when a number of issues are involved. In such cases, it pays to be persistent, but not petty, to insure the best possible deal for your client. In other words, give the agent to understand that you are willing and anxious to bring the matter to a prompt conclusion, you will be reasonable where you feel he is right, but you will not permit your client to pay more tax than you think he is obligated to pay.

Some experienced practitioners, in advance of the audit, classify all items in question into three categories: (1) items they are prepared to concede, (2) items on which they will not give in, (3) "negotiable" issues. Group 1 items will include those issues you feel are hard or impossible to defend, over which the IRS is going to win anyway, and for good measure, you might also throw in a few minor or inconsequential tidbits.

Experienced bargainers feel it is a good psychology to let the opponent have a taste of victory at the beginning. This will convince him that you are a reasonable person, and may put him under some pressure to reciprocate by demonstrating that he too can be reasonable.

It will help if you try to put yourself into the client's shoes so as to better understand what he is up against and what his options are. Until recently, it was an open secret (despite repeated denials by IRS) that agents worked on a "quota" system. In other words, every agent was more or less expected to "raise" a certain amount of money in additional tax assessments, the more efficient he was in that respect, the faster he advanced. Lately, the emphasis on money-raising ability has abated somewhat, and agents are now urged to conduct "quality" examinations even to the point of discovering and correcting errors made by taxpayers in which they overpaid their taxes (but don't rely on it)..

Nevertheless, it is the general consensus by those in the know that agents are still given to understand that they must, to some extent at least, justify the time spent on an audit. In fact, IRS in its annual budget request to Congress usually asks for funds to hire additional auditors on the grounds that each and every dollar spent on their salaries produces X amount of dollars in additional revenues. This means that the longer an agent spends on an examination, the harder he will try to come up with a correspondingly large tax deficiency. It is in your interest, then, to give the agent all the help you can so he may conclude his assignment as fast as possible.

You should also be aware that though all agents are under some pressure to produce money for the Treasury, they are also under pressure to close as many cases as possible. Knowing this can sometimes give you a psychological edge, where all issues have been resolved except for one or two items. As a rule, most agents are anxious to settle such cases to conclude the examination. This is especially true where you have convinced that agent that there is a point beyond which you will not go to reach a settlement.

Here are some additional hints suggested by seasoned tax practitioners:

1. Never hedge when answering an agent's question. Say "yes", "no", or "I don't know". You need not hesitate to admit that you don't know the answer.
2. Never impugn an agent's motives or attack him personally, even if you feel he is being stubborn or arbitrary.

3. If the agent is forced to give in on a hotly contested issue, permit him to do it in a face-saving manner.

4. Do not negotiate in front of outsiders. It will make the agent only more determined to show his competence and bargaining ability.

The Issues Involved

In general, tax controversies fall into two categories. In one, there is a question of law or interpretation; in the other, there is a question of fact. Where the question is one of fact—did the taxpayer really make the expenditures deducted, or did he actually support the claimed dependent; there is no better way to arrive at a quick and favorable solution than by demonstrating to the agent the care with which you prepared the return (assuming that you prepared it). If you followed our previous suggestions about preparing your substantiation at the time the return is made out, your task will be an easy one.

Where the question is one of law or interpretation, you probably explained to the client in preparing the return (again assuming that you did prepare it), that the particular point is open to debate, and it is now up to you to convince the agent that you are right. This will usually call for a certain amount of research on your part, and this is where the tax services and other research and reference materials come in handy, because they cite previous rulings and court decisions on similar or related questions.

Finally, bear in mind that in questions of law or interpretation the agent operates along rather rigidly prescribed rules. He does not really have too much leeway. He may secretly sympathize with your client and feel that a certain deduction should be granted, but if the official IRS position on this item is "no" his hands are tied. A common error committed by many taxpayers (and even some practitioners who should know better), is to try and convince an agent that a certain rule or regulation is unfair. You must understand—and give your client to understand—that the agent does not write the rules; his job is to enforce them, and he has very little say in the matter.

However, where questions of fact are involved, the agent has considerably more authority. He is expected to use his judgment, and as far as the IRS is concerned, if he considers an item substantiated, his decision is generally final.

Preparing the Documentation

In order to substantiate a claimed deduction (this is the area where most problems lie) you must prove (a) that the expenditure was actually made, and/or (b) that it is allowable.

The most common and accepted substantiation of expenditures are canceled checks, but sometimes the agent will want to see the bills or invoices too. In the case of charitable contributions, agents often want to see receipts, especially for larger amounts. If these are not available, you may get a letter from the organization confirming the gift. Bear in mind that for contributions of $250 or more made at one time the client is required to have a receipt on file in any event Casualty and theft losses can often be substantiated by photos, (especially in the case of fire, storm, flood, etc.), newspaper clippings, and police reports (in the case of accidents, thefts, and robberies). If nothing else is available, an affidavit from witnesses or others having knowledge of the circumstances should be submitted.

An excellent means of assembling and keeping control of this substantiation is as follows:

Take a multi-column worksheet and enter the deductions in the first column (or other questioned items) as they were reported on the return. In the second column, enter the amount of substantiation you have available, and in the third column, make a note of the type of substantiation (i.e. checks, receipts, letters, etc.). In the fourth column, enter those items or amounts for which you have no substantiation. Use the remaining columns for notations or explanations as to how you arrived at the unsubstantiated figures. This lets you see at a glance your strengths and weaknesses—what you have in the form of proof and what you still need.

Whether or not you actually want to show the sheet to the agent depends on how well documented the return is. If all or most of the items on the sheet are substantiated, by all means let the agent see how well you did your homework; if not, better keep it out of sight.

Needless to say, this worksheet approach will save you and the agent considerable time, and will also help insure that you or the client do not forget to bring any needed papers or documents. Another advantage of this system is that it helps put you, rather than the agent, in control of the order in which the audit proceeds. As we mentioned before, you can then start with the easy items and gradually work your way towards the tougher ones.

We strongly suggest that before you submit this documentation, you examine it critically yourself, just as the agent would. Check if the documents are convincing and credible. Also check if the date shown corresponds with the return. Too often, a taxpayer will happily furnish a receipt or other document to prove an expenditure only to have the agent discover that the check or receipt shows a different year than the one at issue.

The reason we urge you to examine all proofs and documents with a critical eye is simple. Careful documentation of the questioned items not only makes your case considerable stronger, but also engenders respect for you on the part of the agent. Conversely, too many rejected or weak items will undermine his confidence and may even cast doubt on the entire return.

Incidentally, while preparing the documentation with the client, you have a good opportunity to educate him for the future. Show him how to avoid repetition by proper record keeping, how to pay all deductible items by check, and similar timely precautions. It is unfortunate, but true, that many clients are not impressed with the need for strong documentation until they are faced with an audit.

This is also the time to emotionally prepare the client if the examination isn't going well. If he has resigned himself to the fact that he may be hit with a large deficiency, the shock will be less painful—if and when it comes. And if the audit comes out better than expected, even if he has some additional tax to pay, he will be in a much happier frame of mind. In fact, he may even cheerfully add up all the money he saved.

The Agent's Report

At the audit's conclusion, the agent will write up his report proposing (if you are lucky), either no change, or (if you are even luckier) a decrease in the tax liability shown, or (as in most cases) an increase in the tax liability. The agent is required to explain all proposed adjustments to you or to the client, so don't hesitate to ask about every proposed change. If the agent's findings are agreed to, he writes up an agreement form (usually Form 870) which the client (or you as his authorized representative) signs. Once the agreement form is signed, interest on any additional tax due stops automatically thirty days later. Alternately, the client can stop running up interest immediately by paying the deficiency at once.

If, on the other hand, you don't agree with the agent's proposed adjustments, you can request, and usually obtain, an immediate conference with the agent's supervisor. You may sometimes find the supervisor to be a little more flexible, especially if you are already close to an agreement with the agent. If you do, the same procedure applies.

Warning: If you have a weak case with respect to several issues and the agent gave in on all but one or two (or on a major issue while remaining adamant on minor ones), it may be wiser to settle with him. The supervisor may pick up on those issues

Should You Settle With the Agent?

Suppose you don't agree with the agent or the supervisor? Should you settle anyhow, or is it advisable to strive for a better deal by pursuing the matter further?

The answer to this questions depends primarily on (1) the amount of money involved, (2) the strength of your case, and (3) the client's temperament and willingness to fight (in other words, does he have the stomach for it, even if he has an excellent chance of winning?).

If the amount involved is small, there is little reason to incur the expense, bother, and headache of going through the IRS appeals procedure. You might explain to the client that further appeals are apt to be expensive, time-consuming, and unless the strength of your case and the amount of tax at issue warrants it, he ought to conserve his time and energy by settling. You should also alert the client to the possibility that if he appeals, the return will be reviewed by more skilled and experienced IRS employees, and as a result, additional issues or questions may arise. According to IRS statistics, more than 95% of all examinations are settled with the agent or supervisor, and so only a fraction of these disputes even enter any stage of the appeal procedure.

Nevertheless, if the client, in your opinion, has a good case and the amount involved justifies the bother and expense, further appeals should be considered. The rationale is that there is enough money involved to justify the appeal even if—as is often the case—you are only partially successful. An appeal may also be warranted even if there isn't a lot of money at stake, when you expect to encounter the same problem again and again in future years. In any event, no matter how strong your case is, never assure a client that he is going to win. The most you should permit yourself is to appraise the case as being in his favor.

Going Up the IRS Ladder—The Appeals Procedure

If you and the client are contemplating an appeal, or are undecided, you should of course refuse to sign the agreement form, but instead await the receipt of the official copy of the agent's examination report, accompanied by the "30 day letter". This letter gives you thirty days to either accept the proposed adjustment by signing the agreement form, or initiate an appeal within the IRS. If no response is made the taxpayer will then receive a Statutory Notice of Deficiency or "90 day letter" which means he has 90 days to either pay, or take his case to the Tax Court.

If you decide to appeal within the Service, you should not wait for the "90 day letter", but request—during the period specified in the "30 day letter"—for an Appeals office conference, in accordance with the instructions accompanying the letter.

Remember; once the case reaches the Appeals conference level you can't represent the client unless you are a CPA, attorney, or Enrolled to Practice before the IRS. However, anyone who prepared the return or has knowledge of the pertinent facts, can accompany the client as a "witness".

The Field Examination

The previous discussion relating to office examinations applies to field examinations as well. Since field audits normally encompass a larger area, they usually take longer and require more advance preparation.

When the agent appears at the client's business or home, he should be given a comfortable place to work and be treated politely. There is nothing wrong with offering him an occasional cup of coffee or other refreshments, but don't make him so comfortable that he'll be reluctant to leave and consequently be tempted to prolong the audit.

If lunchtime intervenes and you feel the examination is almost complete, don't offer to take him to lunch or have sandwiches delivered to his desk. An empty stomach may help persuade him to finish sooner and settle the items in dispute on your terms. On the other hand, if he is going to stay all afternoon, you could offer to take him to lunch at a nearby restaurant. If you do, choose a place that is not-too-expensive, and let the agent pay his own bill if he insists, in fact, he is supposed to. Also, don't talk "shop" during lunch unless he broaches the subject first.

Some practitioners like to postpone discussion of the really tough items until near quitting time. This confronts the agent with the choice of having to stay overtime and come home late for dinner (an unattractive prospect, especially since he doesn't get paid extra for it), come back the next day (which may interfere with his schedule); or try to settle with you and conclude the examination. Taking this route, though, involves a gamble, for the agent may decide to come back and spend another half or full day on the examination.

One important point: In case of a field examination, try to ascertain at once whether you are dealing with a regular or a "special" agent. Special agents work for the Criminal Investigation Division of the IRS and are used primarily for fraud investigations. So if a client is contacted by a special agent, the chances are he is being suspected of fraud. Advise the client to hire an experienced attorney immediately and not discuss anything with the agent or disclose any information without first consulting the attorney.

Likewise, if in the course of preparing for the audit, you suspect tax fraud on the part of the client, you should by all means urge him to get an attorney experienced in handling tax fraud cases. If the client asks you to suggest an attorney specializing in this work, you can ask among fellow practitioners or else contact the local bar association. In any event, once the attorney steps into the picture, you should let him handle the case completely, let him decide whether, and to what extent, he will need your services.

Under prior law the attorney-client "privilege" did not extend to accountants and tax practitioners. Thus the IRS, in the course of a client's audit, could subpoena all practitioner-client communications connected with the return. The 1998 IRS Restructuring Act extended the privilege to CPAs and Enrolled Agents, thus protecting the confidentiality of these records, except in the case of criminal investigations.

The procedure to follow if agreement is not reached in a field examination is generally similar to that of an office examination. However, unlike an office examination, a taxpayer seeking an Appeals office conference must file a written protest if the amount in dispute exceeds $2,500. If the amount in dispute is less or in the case of office audits, a written protest is not required but is advisable, especially if difficult questions are involved.

We should also point out that regardless of whether the disagreement results from an office or field examination, the taxpayer can skip the Appeals Conference and go to court immediately. To go to the District Court or Court of Claims, he must pay the deficiency and then sue for a refund. To go to the Tax Court, he simply ignores the 30 day letter, which would then be followed by the 90 day letter. As soon as he receives that letter, he can petition the Tax Court. If he wants to speed the process, he can ask IRS to issue the 90-day letter immediately, and then petition the Court at once.

After your first few IRS audits, you will find that dealing with the IRS has lost its terrors, and can even be quite enjoyable, challenging, and stimulating—not to mention the considerable financial rewards. Furthermore, your success in steering clients through tax audits will prove to be a tremendous client, and prestige builder.

State Tax Audits

While the exact rules and procedures may differ, state income tax audits generally follow a pattern similar to federal tax examinations. State tax auditors too, will be on the lookout for unsubstantiated deductions, unallowable items, and so on. Likewise, the suggestions given regarding conduct with IRS agents apply to state tax personnel as well.

Until recently, state income tax audits, except in the case of larger businesses or well-to-do individuals, were a rare phenomenon. Generally, the states used to rely on the Federal government to bear the brunt of verifying income tax returns. Since taxpayers in just about all states are required to report any changes on the Federal tax return to state tax authorities, the state would eventually get its share of any upward adjustments in an individual's tax return anyway.

During the past few years, however, close coordination has developed between the IRS and the various state tax departments, and as a result, there has been some sharing of the audit load between the state and the Federal government. It is not uncommon to find a taxpayer being audited in some years by the IRS and in other years by his state tax department. Any adjustments made by the state are passed on to the IRS and vice versa.

This means that although state tax rates are considerably lower than Federal rates, it is just as important to fight any upward adjustment on the part of a state auditor, even if the amount involved is not that much. The reason is simply that any increase in the state tax liability automatically results in a much larger increase in the Federal tax.

A Final Word

It cannot be emphasized enough that the groundwork for successfully weathering an examination starts with the preparation of the tax return. You can avoid a lot of grief for your clients and yourself by making sure that every return is:

A. Properly substantiated: Working papers, notes or memos retained with your copy of the return should show what source documents (i.e. checks, receipts, payroll stubs, invoices, etc.) are available to support the return if questioned. If items are based on estimates, make notes as to how the estimate was made.

B. Professionally prepared: The entire return should be neat, error-free, no required data missing, and all required schedules and attachments included.

C. Unusual items explained: Any inconsistencies, unusually high deduction items, and other out-of-the-ordinary entries, that might flag an audit should be explained on the return. At the very least, make a notation on your working papers so that you have an explanation if the return is questioned.

D. Last, but not least, make it a practice to forewarn clients—whose returns show unusual income, or deduction items, or other uncommon features—of a possible examination.

To round out this chapter consider the following additional hints and suggestions:

When preparing for the examination try to anticipate all possible questions and objections the agent may raise. Remember that the agent may have to defend items he resolves in the taxpayer's favor—so help him.

If an audit results in a large tax liability you can generally soften the blow somewhat by arranging for installment payments.

If purely technical questions are involved, an examination may sometimes go smoother without the taxpayer by allowing you to talk directly to the agent on a professional level.

Contrary to what many taxpayers think, last minute filing of tax returns will not avoid a possible audit. The IRS insists that all returns filed receive the same scrutiny, no matter when they are filed.

In taking a client with you to an examination try not to leave him alone with the agent. He may forward some damaging remarks or revelations.

IRS agents will not recommend a tax consultant, but will often express an opinion about his or her qualifications if specifically asked. Obviously, their judgment carries great weight with many taxpayers, so it pays to make sure they have a good opinion of you.

Fees

As we discussed in the section devoted to this subject, fees for representing clients at IRS or state tax examinations, and fees for preparing for such examinations, generally run higher than those charged for routine preparation of tax returns However, our latest information indicates that the differential has leveled off to some extent. While practitioners previously surveyed reported that hourly audit fees ran 25 to 50 percent above hourly tax return fees, that figure had shrunk to under 20 percent by 1997. But part of the decrease may be due to the sizeable increases in tax return fees that took place in the interim.

You should also consider the amount of money involved—and last but not least—your success in the controversy. Obviously, where a large amount of money is at stake, it helps to keep the client informed—to prepare him if he eventually has to pay, or to justify your fee if you succeed in saving him all or even a fair share of the proposed deficiency. By the same token, don't aggravate the sting of an unexpectedly large tax bill with a whopping fee.

In the case of new clients whose return you didn't prepare, when billing them, consider the likelihood of their becoming a regular client. If you charge him less now, you will very likely make the difference up many times over in the years to come.

Incidentally, a 1998 survey, undertaken by the National Society of Accountants, of its own members, disclosed that more than 3 out of 4 do charge an extra fee for representing clients at IRS examinations. Of these, about 86% base their fee on an hourly rate while the remaining 14% charge a minimum fixed fee. Average hourly rates range from $68 to about $105 with a median hourly rate of approximately $84 whereas minimum fixed fees reported ran from $135 to as much as $400 per audit.

Chapter 15
ASSURING YOUR ADVANCEMENT THROUGH PROFESSIONAL SELF DEVELOPMENT

For many years it has been part of my business to indulge in the analysis of businesses and the personalities who operate them. Without one single exception, successful business results from the power released by the positive attitude of mind. With scarcely a single exception, the failures analyzed have been dominated by the negative attitude. I have yet to find an outstanding successful individual who has a negative outlook. I have yet to find a failure who had positive qualities outweighing the negative. I have yet to find even one top executive, man or woman, who doesn't agree with these findings. Success adores the positive.

Douglas Lurton

Very little in this world stands perfectly still. Most everything is in flux—changing, evolving, hopefully progressing, and the ongoing strength of your tax practice lies in your recognition of this basic fact of life. To succeed you will have to meet the ever present challenge—of change.

The changes that concern you will come from two directions: the general tax picture—its rules and regulations—will undergo change, and your clients' needs and circumstances may change from year to year. For your part, you must keep up with these changes—both in your field, and in your clients' lives—not only to maintain, but to advance in your profession as well.

As a tax consultant, you sell your professional knowledge and skills. Your business does not lie in the stack of forms on your shelf, but in the education, training, skills, and competence you have acquired. And like a merchant who must constantly restock, renew, and upgrade his inventory to stay ahead in business, you too must strive to keep up-to-date, refresh your know-how, and sharpen your skills, if you want to get ahead. In fact, due to continuous changes in our economy and concurrent developments in the tax field, keeping up-to-date becomes an absolute must.

To achieve maximum growth of your practice and earnings you need to increase both quantity and quality. In other words,

Most successful tax practitioners find that a substantial portion of their growth in revenues is attributable to their existing client base. In other words, a percentage of your clients will sooner or later open businesses, form partnerships and corporations, enter the professions, start to buy and sell properties, and engage in a host of other financial transactions and commercial ventures that require in-depth tax skills. While you may have the basic knowledge to handle this type of taxpayer, that's true, but it is only through actual practice in the field, through continuous study, through continuous research in handling the various problems that arise—that you become truly proficient in the art of being a tax professional. (More of this later in the Chapter).

Finally, increasing your professional skills will provide you with a high degree of personal satisfaction: your self image and self confidence will expand by leaps and bounds—consistent with your ever expanding reservoir of knowledge, skills, and personal proficiency. You will become a reliable and recognized expert in your field—in a better position to answer to your clients' needs.

Keeping Up-To-Date

As you know, tax law is in a continuous state of flux. Congress enacts new tax rules almost every year, the IRS issues new and revised regulations, the tax forms undergo annual revisions, and the courts keep re-interpreting the statutes. So whether you prepare ten income tax returns a year or a thousand, you owe it to your clients to keep abreast of any changes that may affect them.

Fortunately, keeping current in these matters is a lot easier than you may think, Once you have acquired a clear knowledge and basic understanding of the tax provisions as they stand today, you will have little difficulty following whatever twists and turns those rules may take on in the future. The reason for this is that the basic structure and the fundamental principles do not change. By way of example: the airline pilot does not have to learn how to fly all over again, every time a new plane is on the runway—regardless of how radical the new technology may seem, and in spite of how complex the new gadgetry in the cockpit may appear. Likewise, you will easily learn to cope with any of the changes in tax legislation—now and in the future, because all of it merely extends from a structure you are quite familiar with. But you do have to make a point of keeping up with the changes.

Keeping current on changes in tax provisions—and how they impact upon your clients—remains a 'must'; and there is no excuse for any tax professional who prepares this year's return on the basis of last year's tax law.

If you would like to have detailed, monthly coverage of all important tax developments, we offer—at a nominal fee—our monthly TAX CONSULTANT'S NEWSLETTER. This publication keeps you up-to-date throughout the year on all current legislation (whether enacted, or still in the planning stage), revenue rulings, court decisions, and other tax news of interest. It also contains monthly tax planning features designed to help the practitioner effect maximum savings for his clients and for himself. Although there are a number of good tax newsletters on the market, THE TAX CONSULTANT is the only publication designed especially for the needs of the tax professional who serves the low to middle income client. For further information about the newsletter, please contact the NTTS Publications Department at 1-800-914-8138.

In addition, we suggest you purchase, prior to tax season, one of the annual income tax guides, which are readily available at your local bookstore. NTTS can also provide you with its annually revised U.S. Master Tax Guide, a popular reference handbook of over 400 pages.

Once your practice is well established and profitable, you may want to subscribe to one of the privately published tax services (print, CD-ROM, or Online) which provides a detailed continuously updated compilation of all current tax laws, regulations, rulings, court decisions, etc. Such a service is indispensable to anyone with an extensive practice, but in view of the expense involved (subscriptions range from several hundred to well over a thousand dollars per year), we wouldn't recommend it to the beginner.

How To Sharpen Your Skills And Broaden Your Horizons

If you are genuinely serious about your tax work, you should seek out every opportunity to add to your fund of knowledge and to improve your skills and competence. Any training program, regardless of how thorough it is, can only provide you with the basic foundations you need to enter your field endeavor. Then, once you begin to practice, you will find that your knowledge and expertise will grow with you 'on the job'. In fact, the work itself becomes the best catalyst to assure continued growth. At the same time, however, you can stimulate and accelerate the learning process in other ways as well: by attending classes, by reading, and by keeping in touch with fellow practitioners—especially a mentor.

Most service related businesses offer periodic seminars and classes for those in the field. When you find such offerings in your area (or even nearby) it would be a good idea to attend. For example, the IRS conducts special seminars or workshops for tax practitioners, just make sure the subjects covered in the seminar are in areas of concern to you and your clients. It makes no sense to attend a conference on gift and estate planning (unless you want to possibly go into that area), if all you do is income tax returns for wage earners.

In addition to what you may formally learn, such classes and seminars are also good places to make friends, meet your peers, and exchange constructive ideas. People with more experience than you may offer you valuable tips, you may meet consultants who will back you up—or vice versa—in a moment of professional need, or who may even refer their overflow business to you. Being in the company of those who share the same professional concerns as you can be quite a useful and edifying experience.

Another benefit of attending such gatherings is that you may, in turn, be prepared to share what you have learned with others; to give an up-to-date presentation of your own to clients (and non-clients) on vital, practical, and current tax related topics. Aside from its informational value, this can be an excellent promotional event as well.

Reading pertinent material is an excellent way to grow professionally, but to be most effective it should be handled systematically. Try to designate a specific time of each day, or an amount of time each week (we suggest 2-4 hours) to be devoted to this activity. In fact, try to integrate this into your daily or weekly work schedule. Peruse as much tax related material as you can, but then concentrate on those articles or subjects that are of specific interest to you and your clientele. Also, make sure you read to remember: pay careful attention, underline, make marginal notes—whatever helps you retain the material.

You will get more practical knowledge out of what you read, and retain it all the more if you make a habit of applying what you are learning to particular client situations. If you find an idea that could reduce taxes for a particular client or clients, make a note for yourself to do some further research, and get into it at once. You can well imagine the boost your practice will get if you come up with a few unexpected tax saving ideas for your clients.

Even if an article you read has no immediate practical import, file it away. Try to remember the salient points and keep them in the back of your mind. You never know when you may have an occasion to use this information.

It also helps to read up on tax related material. Therefore, examine the history of taxation, try to remember tax related anecdotes, stories, even cartoons. Your well rounded familiarity with all aspects and facets of the tax world will further enhance your image as a tax professional.

Along with seeking ways of increasing your tax knowledge and competence, make every effort to expand your knowledge and awareness of general current news, general business matters, and economic trends. Stay informed of what goes on in the world-at-large and in the business world by reading journals devoted to such developments. And all of this applies to your employees and associates as well. Their continuous professional growth along these lines will help to enhance their own job performance, and in turn, the reputation of your practice.

The Information Explosion

For better or for worse, we live in an age of 'information overload', whereby a virtual flood of information comes our way daily through the various media; and we—from our side—simply do not have the time or capacity to handle it all. Therefore, in order to make most effective use of the time you have allotted to professional growth, you must constantly screen out—the pertinent from the irrelevant, and the reliable from the false and phony.

- First, identify and locate the best available material, that is worthy of your attention, be it in the area of general news, business news, or tax news. It may exist in books, on disk, on-line, video and audio cassettes, radio, television, or reference services. Then, develop the discipline to discard all that is unworthy of your time and attention. And finally, within the material you have decided to stay with, learn to seek out and focus in on the pertinent facts and figures, and discard the rest.
- With time, try to develop your own media mix. If one good weekly magazine and a local daily paper, or evening radio or TV program is sufficient for your general news intake, then let that be enough. You don't have to read, see, and hear the same thing repeatedly day after day.
- If an occasional tax related text, and a specific weekly or monthly tax newsletter or journal is all you need to keep abreast of necessary tax news, then let that suffice.
- In all the media material that comes your way—from mail, to magazines, to texts and tapes—learn to skim the material, and weed out the waste and redundancies, and home in on the key points.

Getting Into Deeper Tax Areas

The majority of small, independent tax specialists confine their practices to individual income tax returns. This is obviously where the demand is, but there are millions of partnerships, corporations, trusts, and estates that also need competent tax assistance. Needless to say, practitioners who are equipped to handle these more complex tax situations stand to gain a greater financial reward.

If you are currently primarily or exclusively 1040-oriented, it is entirely up to you whether you want to delve into these more complex but lucrative tax areas. If you are content to stay with your practice as it is, there is certainly no need to go into this field. On the other hand, if you do seek broader opportunities and aim for a substantial increase in earnings, then you should consider an advanced training program to familiarize yourself with the taxation and the unique tax problems of partnerships, corporations, trusts, and estates. You can acquire the necessary training through evening courses at a nearby university, or if you prefer to study at home, you can train through the NTTS Higher Course in Federal Taxes. We would be happy to furnish you with full details upon request. Call 1-800-914-8138, or visit our website at www.nattax.com.

Contrary to what you might think, a corporate and partnership clientele is not difficult to obtain even for non-accountants. It's true that most incorporated businesses and partnerships employ year-round accountants, but there are many small business organizations operating in a partnership or corporate form that need few accounting services other than the preparation of their annual income tax and franchise tax returns. In fact, even corporations that do no business must at least file an annual state franchise tax return to keep from being dissolved by the state. And the minimum fee for preparing even simple corporate tax returns can be quite impressive—so it does pay to be versatile in this area. We do suggest, though, that you take it one step at a time: make sure you are fluent in individual tax law and related procedures before you climb up to the next rung of the tax ladder.

The Pre-Tax Season Warm Up

Another way to boost efficiency and increase both the quality and quantity of your output is to spend a few hours—before the tax season begins -to brush up on your tax knowledge. Briefly skim through your reference material, or a current tax guide, paying particular attention to any changes that have occurred since last year—both in the rules and in the annually inflation-indexed amounts (exemptions, standard deduction, minimum filing requirements, phase-outs, etc.) Remember, your clients expect you to be familiar with such routine matters. If there are any commonly encountered areas that you feel weak in, now is the time for corrective action.

Also try to obtain the new tax forms as early as you can, so you have time to become familiar with them—before the season's clients come around. It could be embarrassing to have a client watch you fumble with the mere entry of tax information. And bear in mind too, that reviewing that material and being thoroughly acquainted with the forms gives you a strong psychological edge as you enter the new season. It will increase your self confidence and enable you to convey an encouraging, positive attitude to your clients.

Have You Thought of Specializing?

Everyone knows that a 'specialist' is held in higher esteem and commands greater fees than a 'general practitioner'. This holds true, not only in medicine, law, and engineering, but in tax practice as well. Many practitioners- as their practice and clientele increases—discover that a substantial portion of their comes from specific industries, professions, or businesses. This is only logical since a satisfied client tends to recommend the services of his tax advisor to other friends, associates, and relatives—and very often they are in the same or related line of work.

With time, you will acquire special insight, become more proficient, and 'fine tuned' to the concerns and needs of these specific groups, and of the taxsaving opportunities applicable to them.

In fact, when you get a new client who represents a trade, profession, or business you never handled before, you will find it most helpful to do some research in that particular field, for any special deductions, tax savings or tax problems related to it. If you come up with any tax-saving opportunities, and inform the client about them, you can expect to soon hear from others in that particular line of work.

Developing such 'specialized' expertise will greatly enhance your value and prestige among clients of that group, and the subsequent 'snowball effect' spells more business for you.

We know of tax consultants who—whether by accident or design—have come to specialize in the returns and tax matters of teachers, theatrical personnel, doctors, dentists, and ministers—to name just a few. You can, and should increase your proficiency in lines you already specialize in or would like to specialize in, by becoming better acquainted with that particular trade, business, or profession. And one way to do this is by reading up on the subject, especially trade publications which very often, during tax season, contain tax articles of special interest to that audience.

Finally, if you really want to publicize your interest and expertise in a particular field, there is probably no better way than to publish a tax article in one of these trade journals. It demands research and effort on your part, but you will find it worthwhile. Getting an article into print is not as hard as you might think—especially when you have the financial interest of a particular population in mind. Simply contact the editor of the trade journal, advise him of your expertise in the field, and perhaps submit an outline of your proposed article. It's very likely that the editor will be eager to cooperate, and may even make some valuable suggestions as to the topics to be covered. If your article is well received, you will probably be invited back, and in time you may even become a regular contributor. In the meantime, a clientele representing that line of work will be forming around you.

The Future Growth And Development of Your Practice

If you are a relative newcomer, it may seem a bit far-fetched just now, but it is a good idea to keep the future in mind—even from the start. It's been proven repeatedly that an individual who knows where he is going, and wants to go, is usually the one who gets there. Likewise, even if you are already well established, the fact that you are reading this book shows that you're looking for further growth.

As we have mentioned time and again, the foundation for success in tax work is technical competence coupled with quality client service. Without this it is unlikely that you will get very far. But now, suppose you have personalized, professional service... where do you go from here?

The next logical step is to decide for yourself how far you really want to go. If your ambition is limited to operating a part time seasonal practice that supplies you with a comfortable amount of extra income—fine. You can be perfectly content with this type of practice and derive a great deal of satisfaction from it—as so many continue to do.

On the other hand, if you would like to see your tax practice become a high-volume, high income business, then you should aim toward that goal -no matter how far you still have to go. This means consistently advancing your professional knowledge, missing no opportunity to obtain additional clients, and taking all possible steps to make your office as efficient as possible. A successful, well run, one-person tax practice can yield a very lucrative financial return.

But if you are really ambitious, and anxious to expand, you should set your sights on developing a practice where the bulk of your tax work is done—not by you—but by your assistants. The reason for this is simple: as long as you are paid for your own services only—no matter how well paid you are—there is a built-in limit to the amount of business you can handle. If, on the other hand, you have sufficient volume to warrant hiring assistants, your gross fees and net income will be limited only by the number of clients you can obtain and number of assistants you can properly supervise—or have supervised. Finally, if you feel that you can obtain sufficient clients at other locations, you may reach the point where you will seriously consider setting up additional branch offices. In that case, be sure these branch offices are staffed by competent, personable assistants. Otherwise, you may find your client base declining instead of growing.

The growth of your practice will also depend, in part, on the growth and development potential of your community. A very likely source of clients, at least in the beginning, are taxpayers who recently moved into the area: newlyweds, those entering or opening a new business or profession, and others who have previously not needed—or had—professional tax assistance. Naturally, if you can plug into a community that is on the upswing, you will have a greater pool of potential clients, than if the area is in a state of decline—where residents are moving out.

Another factor to consider in this regard is that a growing, dynamic community undergoes rapid change with more people opening and expanding businesses, bringing in new industry, creating new jobs and positions, etc. And although experience has shown that capable tax practitioners can succeed in a stable or even a declining community—when you step into an area that is 'on the rise', you are very likely to develop right along with it.

Stress Management and the Tax Practitioner

Until now our emphasis has been on 'professional' self development, but we feel it necessary to discuss another type of 'development' as well, that of one's physical and emotional well being. 'Professional' and 'personal' self development really go hand in hand—and one cannot really proceed without the other.

A common concern of anyone involved in today's hectic business world is the matter of emotional, work-related stress, and how to deal with it effectively. To pretend that this type of stress does not exist or to deny that it affects one's work or family life is simply naive.

In a tax practice—or for that matter, in life- it is unrealistic to expect things to run smoothly all the time. They just don't. There are ups and downs, slow and busy times (the peak season carries with it its inevitable stresses and strains), and there's always the possibility of an unexpected turn of events, so you have to be emotionally prepared!

Any business can fail, and for any number of reasons, but one of them—unfortunately—may simply be an 'inability to cope'. In other words: you may know the intricacies and applications of tax law inside and out, and you may be better equipped to handle clients than any other office in town, but if you don't maintain a strong, emotional arsenal within, then you may fail on that account alone.

Therefore it is to your utmost advantage to develop positive living habits and positive personality traits—qualities that generate physical and mental well being. Such habits will serve you well—not only in terms of meeting an April 15th deadline—but they will be of service to you—and a personal asset—all year round.

Here, then, are some practical tips relating to your overall well being that will help you cope with work demands throughout the year. Fortunately, the range of options—and coping mechanisms—is great, and the real key is to establish healthy routines and healthy attitudes.

Exercise and diet are essential features to consider in an effort to maintain emotional harmony. The form of exercise you favor is not as important as engaging in that activity on a regular basis. And whether you like to swim, cycle, walk or run—at least plan to do so regularly. Being physically fit does wonders for your self esteem and increases your capacity for physical and mental work.

Hand in hand with exercise is the habit of eating nutritious meals at regular intervals. Eating foods that provide genuine energy and that strengthen you is essential. Try to avoid the fast-food syndrome, and repeated cups of coffee for quick pick-me-ups and then a shot of alcohol for a calming effect.

Don't isolate yourself. Just as the body needs nurturing through exercise and diet, so does the person's emotional state need nurturing through constructive and satisfying relationships with others. Lacking a partner (or until you get a partner), the sole tax practitioner needs to be aware of this vital point. You need to be able to talk to someone—a spouse or other family member, a mentor, a spiritual advisor—for professional advice and personal development.

It is important that, over time, you create a network of supportive friends and advisors to talk to. Spending time with such individuals on a regular basis can strengthen your knowledge, broaden your perspective, and improve your frame of mind. Without input from others, it's very difficult to know how you are doing and what you should be doing differently. You cannot possibly know all that there is to know, so what you need are people who can give constructive feedback. Finding out how someone else coped in a similar problematic situation, or avoiding the mistakes that others made can be invaluable.

You may be interested to learn that regional 'support groups' have recently formed consisting of sole tax practitioners for the sake of regularly sharing information and advice. Groups meet periodically to discuss mutual problems and concerns, and members of such a group often call each other between meetings to discuss tax matters and questions that arise in the meantime. Such groups have also been instrumental in encouraging and helping new practitioners get off to a good start.

Maybe you can be the first to start such a support group in your area...

Try to pick an appropriate time of year (a slow season) to take a vacation. A vacation that offers you a complete change of setting and pace, and provides freedom from job responsibilities can go a long way to alleviate stress and bring you back into emotional balance.

If you can't take a full vacation then how about a mini-vacation? Sometimes a week-end or even a day away devoted to some favorite activity can do wonders for your state of well being.

Then there are times, to be sure, when you just can't get away at all. But even then there are certain emergency measures you can take to regain your composure. For example, in a tense moment, close your eyes and let your imagination take you to one of your favorite places. Or relax with music. Massage your forehead. Take a nap. Take a walk. Usually a break in activity brings with it a break in attitude as well, and some relief from an agonizing moment.

The important point here is to incorporate into your day slots of time that serve as diversions or relaxations. These 'recesses' can help you reorient and re-energize in a positive and helpful way.

Remember: Given the quick pace and pressures of our day and age, it is not always easy to juggle the many facets of an active tax practice—running the business, maintaining proper relationships, satisfying clients, meeting community obligations, and so forth. But fortunately, there are many good and wholesome ways to create a supportive and helpful surrounding environment. The choices range from exercise to hobbies, from sports and entertainment to meditation and meeting with friends, music, reading—the list can go on. The key, however, is to be in touch with yourself: recognize your needs, your interests, your pleasures, and make time for them, in order to derive the requisite energy or benefit, so that you may continue with your tasks and responsibilities.

Actually, all of this seems to be an age-old rule. In the year 600 BC the philosopher Anacharsis said, "Play—so that you may be serious". And more recently Thomas Fuller remarked, "Refresh that part of yourself that is most weary. If your life is sedentary, exercise the body; if stirring and active, recreate your mind".

The same would seem to hold true today.

Chapter 16

SHOULD YOU CONSIDER A FRANCHISE?

The art of work lies in making your work—you. It is putting the stamp of your unique personality on the work you do. It is pouring your spirit into your task. It is making your work a reflection of your faith, your integrity, your ideals.

William A. Peterson.

Many people think of 'franchising' as something relatively new on the American scene, but it's been around—in some form—for over a century. There are now virtually thousands of franchises in the United States, and as the trend continues to rise we feel that a discussion of the franchise tax field is in order.

The franchising concept is basically a simple one. First, there is a franchisor who owns or possesses rights to a business: its name, its product, its methodology and procedure; and—without relinquishing ownership—he is willing to extend or license the business name and all that goes with it over to someone else, the franchisee—for a fee, and for an ongoing payment called royalty. There are all sorts of conditions attached to this arrangement, but the final legal contract or agreement is termed a 'franchise'.

When you, the franchisee, sign a contract with a franchising company, you become an added unit to that chain of businesses, and they spell out the terms and conditions under which you are permitted to operate the business. In effect, your contract is a license to function under the parent company's trade name and in accordance with their policies and procedures.

For the franchisors the concept is a promising one because it enables them to expand without the large sums of capital needed to form a chain store operation. Furthermore, since each extension is independently owned, the franchisor is dealing with a more motivated-to-succeed group, than if each business operator was a hired manager under his direct supervision.

The main advantage for the franchisee is the array of benefits he stands to gain from the relationship: stepping into an established business with a widely recognized name; uniform products or services, proven, successful methods and procedures—all this tends to reduce the 'risk' factors involved in starting a new business.

Most people are quite familiar with the product-oriented franchises: fast food outlets, ice cream parlors, doughnut shops, convenience stores, hotels and motels—but there is an ever growing number of service-oriented franchises available as well, including a variety of options for the tax practitioner. And, depending upon your circumstances, this may be a viable consideration for you. In any case, you should be aware that the option exists, and of some of the more common advantages and disadvantages inherent to this form of business arrangement.

Advantages of a Franchise:

1. A franchise may serve as a short-cut stepping-stone to your own business. Under this system you have access to a ready made service—complete with image, logo, trade secrets, and regional or national advertising. You pay the parent company an upfront fee and ongoing royalties for the use of its name, specialized services, and assistance with marketing and managing your business. In return you acquire a name and reputation that the public is already familiar with. In other words, you become part of a proven national or regional operation. And—depending upon how well known and reputable a particular tax franchise is—the association could be a definite boost to your career as a tax professional.

2. Conditions vary from one franchise to another, but very often the parent company provides ongoing specialized training, and other forms of service and supplies. In fact, you may receive professional assistance to accompany all phases of operating a business—from location, to equipment, to promotions, and more.

3. There is also the possibility for advanced training and continued assistance from the parent company, with you gaining personally from their experience and proven methods of handling a professional tax practice. The parent organization made an investment in you—you are an extension of their image—and so they are interested in seeing to it that you function well and continue to grow. Therefore, they usually continue to offer you assistance and advice.

Of course, you are contracted to the franchisor for a certain amount of time, but all of this time and experience can be instrumental in preparing you—if you so desire—for a tax practice that's completely yours.

Disadvantages of a Franchise:

1. In exchange for what the parent company has to offer, you do give up a considerable amount of independence. To maintain their image and reputation, most franchisors have to exert strict controls over their franchisees... to which you must submit. There will be areas where you will have very little personal say—for example, fees—and you may feel like someone else's worker rather than your own boss.

2. There will be an initial franchise fee, which may be fairly substantial and in addition, you will have to pay the parent company—either a fixed fee or a percentage of your profits—probably for as long as you are part of that franchise.

3. You are bound by the franchisor's contract terms, and it usually operates in his favor; so that if you want to someday sell out of the business—you cannot do so on your own—but must involve the parent company.

4. Furthermore, whatever faults lie in the system or procedures of a particular franchise—you suffer the consequences. You are dependent upon them to follow through on their promises. If they exercise poor market judgments, you are—more or less—at their mercy. Therefore, you share the burden of a franchisor's faults. Obviously, with a well-established franchise chain your risks in this respect arc minimal you can be pretty sure they know what they are doing, else they wouldn't be around any more.

How to Evaluate a Franchise Before Buying into One:

The Bottom Line

Basically, whether or not to buy into a franchise depends a lot on your personality. If you are looking for an established business name and mode of operation, with most major decisions being made for you, a franchise is definitely something for you to consider.

If you're the independent, go=getter type who doesn't feel comfortable marching to someone else's tune, you may be unhappy with the rules and restraints a franchise entails.

Which carries greater risks? That depends whom you speak to.

According to recently published Small Business Administration (SBA) statistics, franchised businesses have a four-year success rate of approximately 62 percent, while 68 percent of independent start-ups are still in business after four years.

Franchise associations dispute these figures, citing Department of Commerce studies that put the four to five year franchise success rate at 85-90 percent.

The franchise

1. Have a capable lawyer, preferably one experienced in negotiating franchise deals go over every facet of the franchise contract for approval.

2. Find out precisely under what conditions you can terminate the franchise contract and vice versa.

The franchisor

1. Find out how many years this firm has been in business.

2. Does it have a reputation for fairness and success in how it handles franchisees and clientele?

3. Have you seen any certified records and figures reflecting the success and trends of this particular company?

4. Will the parent company assist you with location, training, public relations, capital, credit, and ongoing support system?

You—the franchisee

1. Ask yourself: Are you willing and prepared to give up a measure of independence to conform to this firm's standards and requirements? And do you firmly believe—after examining the particular features of this franchise—that you can work in conjunction with its personnel and policies?

2. Did you adequately survey the territory to know whether a franchised tax service has a market there to accommodate the prices you will have to charge? What other competition already exists in the area?

Other considerations to bear in mind

- When looking into a franchise, proceed with caution and thoroughly examine the franchise you are considering. As with most every business, there are more reputable and less reputable firms to deal with. You don't want to sign yourself over to a firm you will soon regret working for.

  Today, franchise operations are partly regulated by the federal, and in some cases even the state government; and they also have a self policing trade association. Furthermore, franchisors are required to file disclosure statements with the federal government, all of which makes it easier for a prospective franchisee to evaluate each franchise operation.

- One good way to check out a franchisor is to talk with some existing franchisees. Ask them about different facets of their relationship: Has the franchisor done what he said he would do? Have the financial results been worthwhile? Does the franchisor continue to give advice and support? What—if any—are the problems?

You will discover a lot of important information this way, and—even if it means making a number of long distance trips—it will be well worth your time and expense.

- Compare the cost of the franchise you are examining with what it would take to begin a tax practice of your own—including facilities, supplies, promotional activities, and operating expenses. Ask yourself: Would you do better by investing the money in your own independent tax practice—perhaps even running the business out of your own home? This is a question of comparison that is worthy of careful consideration.

  Remember, you already are a trained professional, and—presumably—ready for business. You don't need the franchise for training purposes. Therefore, unless the company is well established and spends a lot on advertising or other forms of promotion to ensure you of a clientele, you may be better off investing your money to promote a practice where you'll be the boss.

- The best time, of course, to evaluate and get the true feel of a tax practice is during the height of the season. If a prospective franchisor has any offices in your area, pay him a visit around this time, you might even go and have a tax return prepared. See, first hand, how they operate, strike up a conversation with clients in the waiting area as you await your turn. Find out what brought them in and ask them how they view the service. Try, if you can, to strike up a conversation with the employee working on your return. You can pick up a lot of worthwhile information about a company's procedures and reputation this way.

In Conclusion

At present, there are a growing number of tax service franchises operating in the marketplace. These range from national organizations to small, regional franchise operations that currently have only a few offices—but are looking to expand with additional franchisees.

If you are seriously interested in a tax franchise, we simply urge you to not part with any money until you have thoroughly examined the franchisor and the franchise agreement, and you—together with your lawyer—are satisfied with what you see.

Make sure you understand all aspects and conditions of the contract: Is it for a few years only, after which you are at the mercy of the franchisor, or is it renewable at your option under terms fixed in the agreement? Are you obligated to do a minimum amount of business per year? Will you have an exclusive territory, and if so, how large is it?

If—after weighing out and comparing all the factors, and examining the agreement, and having done the necessary research—you find that this is a viable option for you, then you are ready to proceed and take advantage of the positive attributes franchising has to offer; for under the right conditions a tax franchise can be a promising venture.

RESOURCES

As of this writing, there are only two sizeable national income tax preparation franchises in existence. Call them if you're interested.

Jackson Hewitt Tax Service (app. 1,250 franchises) 800-227-3278
Triple Check Income Tax Service (280); 816-840-9077

The following firms offer accounting and tax services:

General Business Services (350); 800-877-6181
Padgett Business Services (375); 800-323-7292
Comprehensive Business Services (115); 800-323-9000

There are also a number of regional and local franchise operations. Try the Yellow Pages from a few larger cities in your area, under "Income Tax", or call the International Franchise Association at 202-628-8000. Your local library may also have a copy of the Franchise Annual, which lists over 5,000 franchises plus other valuable information for would-be franchisees.

Bear in mind that with smaller franchises—and especially with recent start-ups—you have to be extra careful.

Chapter 17
HOW TO PROFITABLY SELL OR MERGE YOUR PRACTICE

Presumably, you are either just beginning your practice, or have an established practice and are looking for ways to further develop it. Nevertheless, a book of this sort would not be complete without a chapter on how to terminate a practice—if and when you desire to call it quits.

Or—perhaps instead of selling—you have reached a point in your practice where you are interested in forming a partnership by joining forces with another practitioner.

These are the two options we will be examining more closely in the context of this chapter.

SELLING

Reasons For Selling A Practice

The reasons for selling a practice vary, and this will have a definite bearing on the selling price you can expect. Among the more common reasons for selling a professional practice are the retirement or illness of the practitioner; but other reasons may include planned relocations, the owner's involvement in other business concerns, excessive job related stress, the business no longer appeals, family reasons—or the practice may have gotten too big for the owner to handle, and he just doesn't want to deal with it any more. It is important, however, for you to have a clarity as to why you are selling the practice. And it is also important to note that there are many valid reasons for selling a business, and that 'losing money' does not have to be one of them.

In any case, every prospective buyer will want to know why you are selling; and if it's because you are simply tired of the business, then tell him that. Being honest and straightforward in this regard is the best and surest course to follow.

Your Personal Involvement In The Sale

You should take the sale of your business personally, and not be a detached bystander to the procedure. After all, your practice is your personal achievement—you built it up, nurtured it, and devoted your energy to it—so you should be instrumental in the transfer as well. This is not to say that you should not employ professionals to help you, but they should help you to carry out your goals and objectives—not theirs. You should regularly be in touch with those who are helping you so that they can do so most effectively. Remember, getting a good price will have a profound impact on your own financial future.

Though you should be actively involved in all aspects of the sale, you should bear another important point in mind: naturally, you are emotionally involved and connected to your practice, but you cannot let your emotions and anxieties interfere with the sale of your practice. Don't let your emotions stand in your way. To achieve this you will have to mentally prepare yourself for the negotiation procedure. Be relaxed, be clear and calm; that way, you will be negotiating from strength and not from weakness. Always think positively, this will also help you to negotiate successfully. Don't be critical by focusing on whatever problems you see in the practice, rather, view the situation as one of opportunity. IF there is currently a problem or a setback, it's just temporary and solvable. In other words, your attitude should express the potential that is genuinely there, see the practice as an asset and not a liability. To a great extent negotiation is a tug of wills, and this kind of positive outlook will help you inestimably.

The Role of the Business Broker

Since you may not sell a tax practice more than once in your life, it is not a skill you are likely to be proficient in. Therefore, you might consider placing the practice into the hands of an experienced broker who specializes in the sale of businesses. Buying and selling a business can be a rather technical and complex procedure, so hiring a qualified broker may end up saving you a lot of time, worry, and money as well.

A broker is employed by the seller and earns his commission once the business is sold, it is therefore in his own best interests to be instrumental in obtaining a buyer at a good price. If you are able to find a buyer yourself , and if the transaction is relatively simple, then a good lawyer may be all you need. But if the circumstances are such that the sale is not moving along and it's taking up more time and attention than you can afford, then you may want to enlist the services of a good broker.

A good broker can be of help in a number of ways:

- Since it is their business, they are likely to have a reservoir of potential buyers to draw upon.
- A broker can assist in your making decisions on vital factors—such as evaluating offers, pricing, and terms to consider in the contract.
- A broker can help screen potential buyers and save you time by filtering out the tire kickers from the serious.
- A broker can serve as an intermediary and absorb much of the personal, emotional, interaction that may stand between a buyer and seller, he thereby keeps extraneous material from getting in the way and so keeps the transaction going. With his help the deal is less likely to come undone.

Remember to interview several brokers so you can compare their personalities and their terms, and be sure the one you pick seems capable of the sale of a practice such as yours. Also, ask for references from previous small business clients who are satisfied with the way the broker handled their affairs.

Just as an added note: to hire a broker is an option you will have to weigh out depending on your circumstance, but lawyers are absolutely indispensable when it comes to a business changing hands, and it is essential that you hire a lawyer familiar with the transfer of small business enterprises.

Pricing Your Practice

One of the most difficult aspects to buying or selling a business, especially a 'service enterprise' such as a tax practice, is to determine an equitable and realistic valuation. Appraising a service business is always more complex (than pricing a product enterprise where you have more in the way of fixed assets and inventories) because of all the intangibles involved. There are virtually dozens of factors to examine and consider, by buyer and seller alike, before arriving at a reasonable figure.

It would be most helpful for all concerned if you, as a seller, take the time effort to provide accurate figures, and to break them down so a buyer knows exactly what he is getting.

By way of example this would include:

- **Asset value**—all the tangible assets belonging to the practice such as equipment, furniture, supplies, etc. In setting an asking price you would prepare an inventory and then establish a price for each item. Naturally, you will have to take into account such factors as age and depreciation.
- **Goodwill**—this is a term that is used to describe a conglomerate of intangible features which can account for a substantial portion of the asking price. And not only is this a difficult feature to objectively measure or assign a dollar value

to it, but it is an 'earned' feature that takes years to carefully develop and nurture. 'Goodwill' encompasses a good reputation; a convenient location; the steady, reliable rhythm of repeat clientele; a history of dependable service; a competent staff: name recognition—in other words, the service and spirit of a smooth running machine. On the one hand, how do you assess that; and on the other—this is essentially what your practice is all about!

- Furthermore, every practice must be assessed individually. Like houses, no two are exactly alike. Each practice is accompanied by its own unique circumstances and its own particular strengths and weaknesses.
- Subtle factors such as 'timing' also play a significant role, both in terms of the general economic climate of when you sell, and your own particular 'frame of mind' when you sell. If the country is in the middle of a recession, that will strongly influence the deal, and if you are somehow forced or eager to sell, that too will have an effect on the price.

These considerations are by no means exhaustive—there are more to account for, and some of them are very difficult to quantify, but somehow they must all enter into the overall equation. One thing, however, is certain: as with any other worthwhile endeavor your chances at success increase with the amount of planning and preparation you put into the project. The more you can account for and validate, the better your position is in terms of a successful sale.

The Nuts and Bolts of Selling Your Tax Practice

At the outset you must realize that (in most instances) selling a practice will involve time and energy, and it may take you away from some of the day-to-day operations of your practice. To hopefully make things easier for you, we will discuss the subject by breaking the procedure down into four basic stages. If you decide to employ a broker, he or she will attend to some of these details, or assist you with them.

1. Preparing for the Sale

There are basically two components to this: outwardly it means presenting your business in the best possible light, and inwardly it means getting all your necessary documents in order.

To those who visit and look around, your office and office operations should appear presentable: clean the carpets, paint an ugly wall, make improvements and corrections wherever you can.

At the same time, be ready to document and present your financial operation in a way that makes it easy for prospective buyers to see what the practice is truly worth. You cannot expect anyone to buy a business without a thorough analysis of the past and present financial condition of the practice.

Make sure all your business records are up to date. Have profit and loss statements, income, expenses, balance sheets, and other financial statements available for inspection, along with a copy of your lease—to help the potential buyer ascertain the value of your practice.

It also helps at this stage to prepare a brief written history of your practice—a fact sheet—where you highlight certain positive features: location, traffic, lease conditions, types of clientele, etc.

Another point to consider at this stage: as you prepare for the sale, try to be aware of and sensitive to a potential buyer's frame of mind. Here is someone who is probably looking for financial independence and a chance to grow. To the extent that you understand the buyer's perspective, and to the extent that you can show that your practice can fulfill his expectations—to that extent you will be closer to the final handshake.

2. Locating Potential Buyers

To help you find potential buyers, make a list of those you know who might be interested in such a purchase themselves, or know of others who might be.

Consider professional employees—past and present: If those individuals are technically competent, trustworthy, and willing to stand on their own two feet, then why not sound them out. Such individuals may already be familiar with your clientele and with your office routine, and are logical considerations for the position.

Another likely source would be fellow practitioners who may be looking to move or to expand their practice or, perhaps, to open another office. If you drop a few hints that you are thinking of moving or retiring, and are looking to relinquish your practice, you can rest assured that word will get around fast.

Your list can also extend to suppliers, friends, relatives, or business people in related fields of endeavor—bankers, realtors, accountants, etc.

You can also put classified ads into local and regional newspapers, trade journals, and other special interest publications.

Once you start getting some feedback you face the task of screening your prospects. Early on, try to distinguish viable candidates from those who are just wasting your time. You want to be able to quickly assess whether the prospect can a) afford the price, and b) competently manage the practice. Reasonable and serious prospects will respect and respond favorably to your being candid about this.

As a seller you have every right to request financial and other information to establish whether a potential buyer has sufficient capital and background knowledge to make the business a success. It's time consuming, exhausting, and costly to enter negotiations with someone who is not in a position to follow through. And—you owe it to your clients to make sure that you are putting them into the hands of a capable successor. Therefore, unless you know the individual well, ask for business and personal references, and be sure to check them out. It happens all too often that an unqualified buyer takes over a business and the scenario that follows can be disastrous: if he runs the business into the ground, loses clients, and has to close down—you may be left holding a worthless debt.

As a final bit of advice in this section: try to limit the information you transmit over the phone. As calls come in from prospective buyers, learn not to divulge specific information on the phone. Insist on meeting with prospective buyers in person, so you can size them up.

3. Negotiating the Terms

As a seller you will naturally want the highest price you can get, whereas the buyer will endeavor to spend as little as possible. Negotiation consists of the various stages of bargaining, trading, and compromise between the buyer and the seller. You are both aiming to reach an agreement, an agreement will come through compromise, and compromise calls for taking a logical, clear headed approach toward each element pertinent to the transaction.

It will help you to realize that your negotiating power will be greatly enhanced by how well you know your own financial and the buyer's position. Be aware of all the various bargaining variables that will enter into the discussion. Be precise (write everything down for yourself) about where you can give ground, and where you cannot afford to—and how much you can give. Once you know your own areas of limitations and of flexibility, you will know how and where you can maneuver. That knowledge is power at the negotiating table, as opposed to going there blind, or stumped, or being caught off guard.

As negotiations continue don't rely on your memory or on someone else's to recall important matters. Keep records, make notes, at meetings as well as for phone conversations, then confirm and commit important, agreed upon points to writing. Next, let an attorney draft these matters into legal terminology.

It may also be worthwhile sometimes to have an impartial mediator take part in the negotiation procedure, it may be a small price to pay if the outcome is a mutually favorable transaction. If you agree to having a mediator, try to find someone with broad business experience, preferably one who has previously assisted in business transfers.

Other Points to Bear in Mind During the Negotiations

- How and when the practice will be paid for must also be resolved at an early point in the procedure. It must be clearly defined in the contract, and penalties established in the event of default. As a seller you should take every reasonable precaution to protect yourself and bargain for a safe contract.

- Try to get as much cash as possible 'up front'. If you can get the buyer to pay the whole amount in advance, so much the better; even if you have to go down somewhat on the price. You are more likely, though, to receive 1/3—1/2 down, with the balance due over a period of one or two years. But by no means should you enter into an arrangement whereby the entire purchase price is to be paid out of receipts from the practice. This gives the buyer very little incentive to carry on the business properly, and as a result you may find yourself with no money and no business. If you have a prospective buyer who sounds good but has no money, tell him to get a bank loan, take a mortgage out on his house, borrow money from relatives, or find another method of raising the down payment. Chances are that if he cannot come up with a minimal amount, he will not do well in the business either.

  Very often, contracts make provisions for reduction in the purchase price for fees lost through client drop offs during the first year or two. Naturally, as the seller, you will try to resist such a clause. You can point out to the buyer that just as he will most likely reap the benefits of added business from some of your clients, he will have to assume the risk of others leaving. In any event, try not to make the drop off contingency extend more than one season. Any clients that leave after the first year presumably do so because of their dissatisfaction with the new service provider, and for this you should not be penalized.

- Be prepared to give the buyer substantial help in order to effectuate a smooth transfer of clients; this can increase the value of your practice. If time permits, prepare a fact sheet on each of your better clients, giving essential background information, history, personal idiosyncrasies, etc. Showing a prospective purchaser the pains you have taken and will take to help him through the period of transition and adjustment will facilitate a quicker and more advantageous sale at negotiation time.

4. Closing The Sale

Each and every tax practice is a unique individual enterprise, and there will be a variety of conditions that will uniquely affect the outcome of your transaction. Therefore, a contract that might be satisfactory in one circumstance will not necessarily fit another. The final contract, when it is drawn up, must reflect and protect the special interests and concerns of the buyer and seller and must cover a wide range of contingencies and possible problems that bear upon this particular transaction.

Only a skilled and competent attorney can and should draw up this precise sort of document. Of course, the question always arises as to whose lawyer is going to do this work and how it will be paid for. If both parties agree, you can try to find an impartial attorney who can capably serve both buyer and seller, and then split that fee.

At the closing the sale is completed, contracts are signed, ownership is transferred, and preparations are made to transfer the practice to the new owner. The buyer now stands on the threshold of a new venture, while the seller will either embark upon his own new endeavor, or perhaps enter a more leisurely stage in life. But in any case, the most successful transactions are those where continued cooperation and communication exist between buyer and seller, especially in the crucial first few months after the sale. The buyer should feel that his inquiries are welcome and that advice and assistance will be freely given, and the seller, by being available to help and advise, stands to protect the future payments due to him under the terms of the contract.

MERGING OR RESTRUCTURING

Most tax practices start as sole ownerships—whether proprietorships or single owner corporations—and remain that way until the business has grown enough to warrant considering conversion into a partnership or multiple owner corporation.

The Legal Entity

Whatever legal form of ownership you operate under (sole proprietorship, partnership, S corp., C corp., PC, LLC), you should realize that each type of entity has distinct drawbacks as well as advantages, both tax- and business-wise. So before switching from one form to another make sure to consider all pros and cons. The key here is to select the form of ownership—utilizing professional advice, if necessary— that best meets your specific individual needs.

But legalities aside, there are only two basic operating structures: sole ownership and multiple or shared ownership.

Sole Ownership

The sole ownership is the most common way of starting out in a business. As a sole owner you are totally responsible for your own tax practice: you make all the decisions, and you answer only to yourself. You open the office and lock up, you set the policies, you hire and fire, and you keep all the records. A sole ownership offers you the greatest psychological freedom and reward, but then the entire burden rests on your shoulders as well.

Multiple Ownerships

The traditional multiple ownership entity in a professional service business is a partnership, though many now use any of the other above mentioned forms. We will therefore employ the 'partnership' designation in our discussion.

A partnership is an association of two or more people who together own and operate a business. There are basically two ways of forming a professional partnership:

A. By merging with another independent practitioner; in other words—combining two individual practices, or

B. By taking an employee or other suitable individual into your business, and thereby converting your practice into a partnership.

The idea of having a partner in your business can be very appealing, it means having someone to work together with, rely upon, and share ideas with. But before you proceed with such an arrangement you should be aware of all the foreseeable advantages and disadvantages, and then—after careful consideration—you will hopefully be in a position to make the right decision.

Operating a professional practice in partnership form certainly has its advantages, as evidenced by the fact that so many accounting, legal, medical, and other professional services operate in that mode. One prime advantage is that it avoids the dangers inherent in a one person operation. An organization run by two or more individuals can generally cope with a brief or even prolonged absence by one of the principals—due to vacations, illness, physical infirmity, or death. The latter contingency, in particular, should not be overlooked. Where a sole practitioner is suddenly incapacitated or dies, his family generally has no choice but to dispose of the business immediately, and at a distress price. Otherwise the practice will soon be worth nothing. In a partnership, however, arrangements can be made to have the surviving partner take over based on predetermined conditions.

A partnership also fosters the pooling of talents and specialties of each individual partner, which gives the overall practice greater strength and diversity. One partner may be more of an introvert, the other outgoing; together they can divide the work according to their skills, talents, and preference—thereby creating a more effective practice. Or one partner may be good at marketing while the other is better at organizing.

Also consider the reduction in overhead—one office instead of two—and the possible reduction of outside help. And if you are growth minded, it is usually easier to attract top quality assistants to a partnership.

Another advantage is that you will likely attract a greater variety of clients: some may sync better with your partner, and others may work better with you.

But before you rush into drawing up a partnership agreement, you should be fully aware of the disadvantages as well.

First, if you are used to being fully independent, you may recoil at the built in restraints: you will not be able to run everything your way; you now have another person to account to—for everything. More importantly, there may be differences in temperament, personality, or philosophy that may prove difficult to reconcile. For instance, if your goal is to develop a low volume but high quality tax practice based on personal service to your clients and your prospective partner has visions of a mass volume 'assembly-line' approach—then you may be heading for real trouble. Your clients, in turn, may be apprehensive about your service losing its personal touch. It doesn't matter whether these fears are real or imagined, either way they can harm your business.

There is also the danger that small disagreements between partners may escalate into a major feud, and even if they do not, a negative undercurrent can eventually undermine the genuine effectiveness of your collaboration.

Nevertheless, if after carefully weighing the pros and cons, you feel you have the right person to partner with, then by all means go ahead.

In forming a partnership it is important to get a good attorney—preferably one experienced in professional partnership contracts. Many potential sources of aggravation can be eliminated by drawing up a detailed agreement that spells out clearly the rights, duties, and privileges of each partner, and leaves as little as possible to misinterpretation. Most important: make sure provisions are made for an equitable dissolution of the partnership, if—for some reason—it doesn't work out. We also recommend that the agreement stipulates that any disputes be resolved through binding arbitration. We don't mean to appear pessimistic, but this may someday save you a bundle in legal fees and court costs.

The Financial End of the Partnership Agreement

One of the more critical clauses in a partnership agreement is the interest or equity assigned to each partner. In the case of a merger of two individual practitioners, both having practices of approximately equal size; presumably, both partners will receive the same interest in the partnership. But if one practice is larger, more lucrative, or otherwise more desirable than the other, that partner is normally entitled to fair compensation for his extra contribution. The same holds true where one partner contributes considerably more in physical facilities and supplies.

An alternate arrangement would be for one partner to make a cash payment to the other for the agreed upon value of this extra contribution, and then have both start as equal partners. To estimate the comparative value of each partner's contribution, you can apply the criteria we discussed for establishing the sale price of a practice. In other words, figure out a fair selling price for each practice, and then adjust the partnership equity accordingly.

Not to be overlooked in the agreement are salary, or drawing arrangements. Will both partners draw equal pay? Will one devote more time to the practice than the other? Or are there other reasons that justify different salaries for each partner? How is outside income to be handled? What about expense reimbursement? All such questions should be settled beforehand or they may come back to haunt you later.

Another essential part of the agreement (which we already touched on) is an equitable provision for the remaining partner's acquisition of the practice in the event of the other partner's death or incapacitation.

Here, too, it is important to establish—before entering into a partnership—a method of evaluating the total worth of the business, such as 125% of the previous year's gross fees. Assume for instance, that at the time of a partner's death, gross fees of a two man partnership amount to $120,000. The total assigned value of the partnership would then be $150,000—$180,000, and each partner's share (assuming a 50/50 arrangement) would come to $75,000—$90,000.

Naturally, provision should also be made as to how, and under what terms, this price is to be paid by the partner taking over the business. A common arrangement is to take out a life insurance policy, with a face value equal to the approximate purchase price, on the life of both partners (or often the senior partner) with the proceeds earmarked for purchase of the deceased partner's equity.

In any event, no matter what the arrangements are, they should be agreed upon beforehand, and spelled out clearly in the partnership contract.

Taking An Employee as a Partner

It often occurs that an employee performs so loyally and so well, that the employer offers him a chance to join as a partner. Or—a practitioner who would like to retire may find that the most logical choice for a successor to the business is one of his own staff members. This makes good sense, for here is someone familiar with the clients as well as with the entire office procedure.

The steps for evaluating a practice have already been outlined. However, one problem often encountered in negotiating a buy-in agreement are the terms of payment. One common method is to let the new partner acquire the interest in the partnership—and pay for it—in agreed upon stages. In other words, he might purchase 10—25% in the first year, another 10—25% the following year, and so on, until he has achieved full partnership status. This type of arrangement also makes it easier to dissolve the partnership early on if it does not work out.

Here too, as we pointed out before, it is important that the incoming partner put down some cash, so he will have a decided stake in the business, and the required incentive to put his best into it.

One thing, however, to be aware of is that good employees do not necessarily make good principals. Just make sure you have a good escape clause, and use it at the first sign of serious trouble.

Dissolving a Partnership

No matter how well matched the partners appear to be, some partnerships—like some marriages—just don't work out, and the sooner that fact is realized, the better.

If the business was formed by the merger of two or more individual practices, the accepted method of dissolving it is to have each partner take with him all the clients he brought in. New clients obtained by the partnership are generally divided equally among all partners on the basis of gross fees. In some cases, however, each new client is assigned to the particular partner who either acquired him or worked with him most, or all of the time.

If the partnership was formed through a buy-in arrangement with an employee, or other individual who did not have an established practice, you have two choices. If you brought the new individual in with a view towards later retirement, you can opt for early retirement and give him an opportunity to buy the entire practice from you. If you are not ready to retire, or the employee does not have the means to acquire the practice (or you feel he can't handle it) then you will simply have to refund his investment.

It cannot be stressed often enough that all these contingencies should be spelled out clearly in the original partnership agreement.

Choosing Business Partners: A Few Tips

1. Generally speaking, look for a partner who supplies you with a talent, skill, strength or expertise that you are lacking. Don't just duplicate common strengths unless the one goal you share is to simply sustain a greater workload.

2. Make sure that you and your prospective partner share similar ideas about business goals and methods of how a business should be run. Furthermore, investigate the work habits of the person under consideration. Openly discuss and

define what you think your respective roles should be. Never accept what others or what a prospective partner says at face value.

3. Be wary of an authoritarian figure who is only really happy when he 'runs the show'. What you really need in a partnership is someone who is tolerant, compatible, respects others, appreciates the benefits of teamwork, and is someone you can trust.

4. Be cautious of prospective partners who have a lifestyle radically different from yours, whose habits may be excessive or extreme, are big spenders, or whose value system is not in consonance with yours.

5. Do not pick a partner for monetary reasons alone, i.e. because you are financially desperate and he can improve your situation with his input of money. After the money is gone, you may be stuck with a partner you can't live with. Only pick a partner for the right reasons.

6. Be especially careful before choosing family and friends as business partners. If it works out it can be an excellent arrangement. But if not, you may be faced with the unpleasant choice between destroying a relationship and hurting your business. In fact, the same holds true when considering a prospective employee. The old dictum "Never hire anyone you can't fire" is still valid, as many have learned from bitter experience.

Chapter 18

SPECIAL PRACTICE BUILDER REPORTS

As a final chapter to this volume we are reprinting several 'Special Reports' that we send to our graduates summarizing certain suggestions and hints to help them successfully start and operate their tax practices.

Although a number of these ideas and strategies were already discussed in detail in earlier chapters, these reports will be helpful, both for the additional tips and techniques they contain as well as to serve as a convenient review of some of the major points emerging from Building, Marketing & Operating a Profitable Tax Practice.

REPORT #1

Professional Ways to Advertise Your Service

The tax profession—being a service—is not like other ordinary business enterprises. More like a doctor or a lawyer, for example, you deal with some of the most confidential areas of a person's life. It falls into a category of information still regarded as 'strictly private', placing you in a privileged position of trust. Public attitudes toward you, therefore, require that you conduct yourself and your practice in a conservative manner—that you stay within the traditional bounds of propriety and follow an accepted code of professional ethics. And this applies especially to the methods you use to promote your service.

At first glance, such restrictions may appear to limit you, but there is plenty of room for effective, business-building, promotional activity Just bear in mind—that aside from any advertising effort—your professional reputation really hinges on your ability to provide a competent, qualified service. Don't lose sight of that: when you do a good, conscientious job for a client, you sow seeds that will bear fruit in new clients coming to you by recommendation. Word-of-mouth advertising alone accounts for the extraordinary success of many tax practitioners. But the growth of your business entirely by this means, though virtually certain, would be slow. Here then, are some simple, tested ways to get off to a good start and accelerate the process of growth and increased profits:

Personal Exposure—Become a joiner of local business, social, church organizations. Carry business cards. Conversational opportunities to let people know what you do part-time or full-time will bring contacts with prospective clients. Relatives, friends, and business associates will be eager to help.

Business Exposure—The following devices constitute a complete advertising and promotion campaign. You may use them all, or just those that best fit your circumstances.

A. If you live in a private home, erect a TAX CONSULTANT "shingle" with your name and phone number so passers-by can see it. Make it dignified. If you have an apartment, include TAX CONSULTANT on your name plate on the building directory.

B. Whether your office is at home or out-of-your home make it comfortable but businesslike, with tax and other reference works, small file case, your framed NTTS certificate and other fraternal, business, or membership certificates in plain sight. A client's first impression of you is important, so do whatever you can to create the proper image—that of a professional who knows his business.

C. Have self-adhesive stickers made showing your name, business, address, and phone number. Stick one on each duplicate copy of a return you give to a client for his personal records. Repeat business is what you want, and the sticker serves as a reminder.

D. Develop a series of short, friendly letters to take advantage of business-building opportunities:

1. To New Residents—a letter of welcome and an offer of professional tax advice they may need.
2. To Newlyweds—a letter of congratulations and offer to help with tax problems that come with their new change in tax status.
3. To New Parents—a letter of congratulations along with an offer to take a new look at their income taxes—with your hope to be of assistance.
4. To New Businesses—a letter offering your best wishes for success, and information about your qualifications to service their tax matters.
5. To selected prospects—a letter emphasizing the personal aspects

of your tax service as opposed to the trend toward impersonal service, with a request for an opportunity to be of such help.

E. List your service in the Yellow Pages under "Tax Return Preparation".

F. Contact your local newspaper editor about supplying material for a column on TAX FACTS AND TIPS with your name as the contributor. You don't have to be an accomplished writer. A rewrite person will usually put your material into suitable form.

G. Place a classified or small display advertisement in your local paper. Limit your message to a simple announcement, such as "PROFESSIONAL TAX CONSULTING SERVICE, Personal and Business Income Taxes, Payroll Taxes, Tax Planning." Follow with your name, address and phone number. Do this on a regular basis during the tax season, on an occasional basis between seasons.

H. If you have extra time, consider offering your services to help out accountants, tax attorneys, and other tax specialists during the busy season. Some of our new graduates add several thousand dollars this way to income earned from their own practice.

There are other ways a Tax Consultant can advertise his service in good taste, but as with all your efforts, you will have to determine whether the results are worth the effort and expenditures.

Whatever method you use, be sure to keep one thought uppermost in your mind: it must reflect and be consistent with the dignity and professional ethics expected of those who serve the needs of taxpayers.

REPORT #2

How to Cope with Chain Store Competition

Here's a letter we recently received:

> "I have a disturbing problem which I would like to discuss with you.
>
> "The past three tax seasons I have operated a part time tax service from my home. My clients have steadily increased and prospects, for a continuing, growing business seemed good.
>
> "However, I'm greatly disturbed this year because rumor has it that a nationally advertised tax service will soon open a branch office in my town. Once this happens, I fear that I will lose all or most of my clients. Is there anything I can do to compete with this type of service?"

Because of the nature of the problem and it's frequency, we will devote this report to a discussion of how to deal with this type of competition.

When faced by a large multi-office competitor, the independent tax practitioner must carefully analyze his own, as well as his competitor's, strengths and weaknesses. Then, instead of panicking, he can decide on how to effectively challenge the competition.

For example, chain tax services usually feature, or advertise, quick, while-you-wait service, "low fees", and mass production operation. Sometimes they also guarantee to pay any interest and penalties (but not additional tax) arising out of errors made by them in preparing a return. To procure clients they undertake considerable local and national advertising campaigns. And in order to service a large number of anticipated clients, they will quickly train numerous assistants and clerical help. After the season, they generally close up shop, leaving no more than a care-taker number where someone connected with the organization can be reached. In larger cities where they have a number of locations, they'll keep open year 'round and close down the rest.

This, with some variations, generally describes the kind of tax service rendered by the typical chain organization.

In contrast to this, the Independent Tax Preparer—like the writer of the above letter—offers his clients concerned, personal attention, expert, unhurried, tax advice; and skilled tax return preparation in a professional setting. Frequently, he will even come to the client's home or place of business. Although he may operate his office only during the tax season, he is a full time member of his community and is available year-round for consultation or for assistance if a client's return is audited or questioned.

So now, where do you go from here?

First, we suggest that you do not try to out-advertise the competition. No matter what your advertising budget is, these organizations pour tens of thousands of dollars into sustained advertising campaigns, which is considerably more than the average independent tax practitioner can afford.

Nor do we think it is good policy, or good business, to cut your fees in order to engage in a 'price war'. Incidentally, it is important to note that in most cases the taxpayer whose return is prepared by a so-called 'cut-rate' chain tax service does not pay the low advertised price... but rather a considerably higher amount. In fact, a while ago a recent "Business Week" survey found that the average fee received by one large tax service organization was double their popularly advertised price.

After eliminating the areas where you cannot successfully compete, let us concentrate on your strong points, and those areas where you can easily and decisively 'lick' your competition. And these areas are:

- Quality professional service,

- Personal attention,
- Client loyalty,
- and, perhaps of greatest significance... a high-grade clientele.

We are fully aware that there will always be plenty of individuals who react only to price, and to whom quality service means little, if anything. But we are equally aware that there is another group who both need and appreciate high-quality tax service. It is this second group of clients and prospective clients that can help you build a successful, rewarding, and highly satisfying tax practice.

Bear in mind too, that the first group generally consists of lower income taxpayers who pose few tax problems and provide little opportunity for tax savings (and lucrative fees), whereas the second group usually consists of men and women who are better educated, earn larger incomes,—often from a variety of sources—and as a result require more knowledgeable (and more profitable) tax preparation and advice. In essence, then, it pays to focus your attention on this group of taxpayers. This is the group least likely to be lured away by a mass-production, assembly-line tax service.

Moreover, many independent tax practitioners find they are able to attract and retain a sizable following even among low-income, simple-return, taxpayers in the face of stiff competition from the large tax-service organizations. And the reason for this is simple: In this age of increasing automation, people desperately yearn to be treated with concern and as unique individuals. They appreciate, and remember, a sincere smile, a personal touch, a friendly word, a flicker of recognition.

By contrast, the taxpayer who submits to the typical 'chain store' type of service may get efficient, even courteous treatment, but it's swift and express lane style. To the employee who prepares the return, this taxpayer is just 'another face in the crowd', and one that he will very likely never see—or service again.

No wonder then that the wise practitioner who capitalizes on this inherent weakness in his competition has little fear of being pushed out of business by the giant tax-service organizations. Instead, he can look forward to steadily increasing recognition, advancement, and income, sparked to a large extent by a growing number of satisfied, loyal clients.

REPORT #3

Should you "Guarantee" Your Income Tax Returns?

Our advice is often sought as to whether or not it is advisable for practitioners to include a 'guarantee' with the income tax returns they prepare. Some of the larger tax services, you may be aware, advertise that all their returns are 'guaranteed'.

Before answering this question, let us first examine the 'guarantee'. Under the terms of a typical guarantee, the tax service undertakes to reimburse the client for any additional interest or penalty incurred as a result of their error or negligence. Note that the service does not undertake to pay any tax deficiency that may be levied against the client, nor does the guarantee provide that the tax service will represent the client (with or without fee) before the IRS and argue his case in the event of an audit.

Also note that the guarantee is strictly limited to interest and penalties incurred as a result of the tax service's or tax consultant's error or negligence. Thus, if the error or penalty results from the client's failure to disclose information or from his inability to substantiate deductions claimed on the return, the tax preparer is absolved of any responsibility.

Instances where penalties are levied against a taxpayer because of the tax preparer's error or negligence are extremely rare (except in the case of obviously fraudulent 'tax consultants'), and interest payable for the same reason—while perhaps a little more frequent—is usually a negligible amount; so one may wonder if this so-called "guarantee" is much more than an advertising gimmick. It does not seem to carry much practical value to the client, nor does it carry much risk to the practitioner.

For this reason, if you are in competition with tax services offering this type of guarantee, and if clients ask you about it, we see no reason why you should not offer the same to your clients (orally or in writing).

There is another sort of guarantee some practitioners give, under which they promise to represent the client without charge before the IRS in the event of an audit. Note, they will not pay any additional tax due, they merely offer their services free of charge.

This type of guarantee simply boils down to a matter of dollars and cents as far as the practitioner is concerned. If you have an established practice, you can reasonably be certain that a number of your returns will be audited, the question is merely how many, and how much work on your part the audits will entail. The guarantee then is bound to cost you a certain amount of time, plus a loss of fees you otherwise would collect for representing these clients. This cost must then be balanced against the benefit you derive from it, that is—the amount of additional business it brings you.

In our opinion, the benefits are not worth the cost. In fact, we feel this type of guarantee has a distinct drawback to the client as well, because it puts a certain amount of pressure on the practitioner to 'play it safe' side by resolving all questionable income and deduction items in the government's favor, rather than in the client's. This is generally not in the client's best interest, for though it may be advisable to avoid controversy on minor items, there are many times when the amount of money involved justifies taking a chance on an audit (or even incurring a certain audit)—if you feel you stand a reasonable chance of substantiating your position. If all or most of your competitors do guarantee no-charge audit representation, you have two choices: (1) go along with the trend, or (2) set yourself apart from the competition by refusing to engage in a practice that, you are convinced, is not in the clients' best interest. You can be sure that the majority of clients and prospects—especially the more intelligent, more desirable kind—will understand and accept a clear, sincere, explanation of your reasons.

REPORT #4

How to Develop Tax Business During the Off Season

For most practitioners the season begins about mid-January and ends April 15th. Here then is a novel and effective way to extend the tax season another few months. Reports indicate that this plan is not only a lucrative one, but more important, a rich source for new clients.

The idea is quite simple. During the tax season—especially during the last few hectic weeks—many taxpayers have their returns prepared by run-of-the-mill tax preparers who are either not fully qualified, or do not take the time to do the professional job a tax return deserves. When examining a new client's previous year's returns, you have probably often found errors, inaccuracies, overlooked deductions, or other missed tax saving opportunities. In fact, some tax specialists routinely examine previous tax returns of new clients with a view towards filing amended returns and obtaining refunds for them. Unfortunately however, during the busy season, the average practitioner has his hands full keeping up with current returns and does not have the time or patience to delve into tax affairs of previous years.

A logical solution would be to suggest to new clients, at the time of the initial interview, the possibility of examining previous returns, and then set up an appointment for after the tax season to pursue the matter. Then send the client a reminder a week or two before the scheduled appointment.

You will also find it worthwhile to review the files of those taxpayers who had complicated returns, usually those who enjoy a higher income. By spending some time with each of these returns, you should be able to find possibilities and opportunities for instituting tax saving devices for the future: perhaps some kind of trust fund for charity purposes; or changing the business arrangement from an unincorporated business to a corporation, or to a family partnership; or transferring some property from parent to child. There are many opportunities which you probably could not consider during the tax season. But now, by spending several quiet evenings with these returns, you could probably develop a whole line of worthwhile suggestions. And what could be more pleasing to a taxpayer than to receive a note from his Tax Preparer offering advice on how to save money on last year's return, or profit from new tax angles the coming year.

Suppose you carry this idea one step further, and send a letter to selected (non-client) taxpayers in your area offering to thoroughly review last year's return at no charge. If your examination turns up any areas for refund or tax saving opportunities, you will so advise the client and—at his option—file an amended return. Needless to say, any client for whom it pays to file an amended return will not only be happy to pay your fee, but will be your client for keeps. Moreover, many clients (even if you did not discover a sufficient amount of tax savings to make the filing of an amended return worthwhile) will be impressed and will give their future business to you.

Now if any of this sounds strange to you, be assured that it has been done, and is being done, by an increasing number of aggressive tax practitioners—with results beyond their expectations. They report that, on the average, one out of every three or four returns examined, reveals substantial errors, or omissions.

But what happens if, instead of discovering an overpayment you find that the taxpayer underpaid his tax—either through claiming non-allowable deductions, or making an error on the tax computation? Surprisingly, many clients are grateful for this information, which may avert an IRS deficiency notice or even an audit. If, however, the client refuses to take action to rectify the underpayment, you of course, are under no further obligation.

We advise you to give this matter serious consideration. You may even come up with some ideas of your own;, if so, let us hear from you.

A graduate recently sent us a sample of a short, simple, ad that he runs in local church bulletins, house, trade, and organizational journals. In spite of the limited circulation of these publications, he reports excellent results and find that the work which comes in 'fills the gap between the seasons'.

The ad reads as follows:

Did you take full advantage of all your allowable tax deductions and tax-saving possibilities? An amended INCOME TAX RETURN FOR 19__ may yield a substantial refund now! Call 000-0000 for free consultation.

Incidentally, after taking our advice, and preparing a number of such amended returns, why not write a short article citing case histories (omitting names of course); detailing the errors or omissions you found, what you did about them and citing the amount of taxes saved or refunds obtained. You can use it as a mailing piece and/or as a newspaper column. It will make for interesting reading, and should help to bring in new business as well.

Remember, many tax preparers have discovered that servicing clients after the tax season can very often be as profitable as in-season tax preparation—if not more so.

REPORT #5

More Post-Season Building Ideas

A problem that virtually every professional or business person faces, and bemoans, is the lack of time to "think". Because of the daily rush and pressures, little time, if any, is spend on quiet thinking and planning.

But with the April 15th deadline safely behind you, and extensions taken care of, why not spend some time in creative thought on how you can increase your client base as well as on how to develop new avenues of service to present clients. If you adopt one or two of the strategies we discuss below, you will be taking a tangible step towards building a more successful tax practice.

Do Something Extra To Keep—And Win—Clients

The late Orville Reed, a famous direct-mail expert, once told of an automobile dealer who built up an unusually big business and attributed a great deal of his success to one simple strategy.

A few weeks after a customer traded in his old car for a new one, he mailed him a $25 check with a letter saying that he was able to sell the old car at a higher price than he thought, and that he feels it only right to share this unexpected, but welcome result with the previous owner.

Such a letter not only builds good will but the customer will undoubtedly tell his friends about this and the dealer will go on to get priceless word-of-mouth advertising.

This potent idea is adaptable to many situations because everyone likes to receive an unexpected something extra. For example:

a. Looking over your client list of last year it occurs to you that Mr. Smith's son just became engaged. Write a short note to congratulate the father and mention the fact that Junior's up-and-coming marriage may carry some tax consequences for him, then invite him in for a discussion about the matter. Such a response on your part will help Mr. Smith become a lifelong client of yours and a staunch booster of your services.

b. Mr. Jones is building an extension to his home or investing a considerable amount of money in landscaping and alterations. A brief reminder that keeping accurate records of his expenses will save him both taxes and headaches in the future, will convince Mr. Jones that you are not just a "tax return preparer" but a professional who takes personal interest in his clients.

c. Last year you told Mrs. Thomas that she has much to gain by keeping detailed records in connection with her part-time home business. But since we all tend to forget, it may be wise to check in with Mrs. Thomas, by letter or by phone to see if she is following your advice and thus getting the full benefit of your services.

There is no doubt that if you spend just a few minutes thinking about each client you will come up with a long list of such valuable 'goodwill' builders. Remember that even your best, most conscientious service during the tax season may often be taken for granted, but the practitioner who takes the time and trouble to think about a client and his tax problems after the tax season, will make a most favorable and long-lasting impression.

Think Ahead

The previous discussion was devoted to improving the 'quality' of your service. But what about the many promotional ideas you may think of during the tax season... when you have no time to do anything about them?

During the season, you are bound out of necessity, to a strict regimen which does not allow you to think far beyond your current commitments. But the months after the tax rush provide you with an excellent opportunity to evaluate the season's results, in terms of the number of taxpayers serviced, amount of income per time invested, the value of advertising, etc.

Also, consider the effect of local competition on your business. Where is the market going? Would it be to your advantage to join forces with others, or branch out by yourself? Did you lose any clients? And if you did, why? Now is the time to creatively consider these various issues.

Getting lists of new prospects, preparing mailings, making contacts with groups of prospective clients, seeking and obtaining publicity... and many other ideas could and should be worked out now, when you have the time and peace of mind to devote to these concerns.

How about training an assistant in simple office and tax procedures? How about beginning to look for a more suitable location for your next year's tax office? This is the sort of speculation which, once started, may direct you to a greater and more lucrative practice.

Remember that next January and February you will again not have much opportunity to grow, because at that time you are too concerned with your immediate problems. Now—between one tax season and the next—is the time to assess and plan for your future growth.

Thank You Letters

Now, with the busy season behind you, why not send a friendly letter of thanks to each of your clients?

Furthermore, since client referrals are the life-blood of almost every professional practice, what would be more appropriate than to send a special thank you note to each and every client who referred a friend or acquaintance to you this season? This is another thoughtful gesture that will pay dividends in goodwill and further increase referrals in the years to come.

REPORT #6

Attaining Success

It is interesting to note that although most graduates of our training program begin their practices with the same basic knowledge, the results over the years show that some will attain the ultimate in business success while others will establish a mediocre practice. And this just about parallels every situation in life: the most successful person does not become so due to superior knowledge, "unusual breaks", special contacts, or a streak of good luck, but rather on account of two factors—the real keys to success:

a. Direction: Know where you're going.

b. Determination: Making up your mind to get there.

You have undoubtedly read biographies of successful people. In reading these, take note that the real turning point in a person's life comes following a clear awareness of one's precise ambition in life. Whether it was laying the first trans-Atlantic cable, building the longest suspension bridge, or heading a multi-billion dollar corporation, whatever the field of endeavor—the steady climb to success begins once you establish a clear cut goal.

Once a person is determined to succeed in a specific area, the next step is one of perseverance: to maintain the drive and motivation to achieve that goal—in spite of obstacles along the way.

How can we apply these principles in practice to building a successful tax practice?

First, when we speak in terms of a successful tax practice, it is important to remember that the word 'success' has different meanings to different people. For example, we once received a letter from a graduate in California. He wrote that over the past three years he was able to open three offices. He now employs ten secretaries and three tax preparers, and hopes to succeed in becoming the "largest tax consulting firm in the West". On the other hand, we hear from former students who annually prepare anywhere from 25-50 tax returns and, as they put it, "have attained success" and are "perfectly happy".

So the first thing we must do in our own minds is define for ourselves exactly what we mean by success. The person who envisions opening a new office in the most fashionable part of town, is thinking, even now, in different terms than the one whose concept of success is limited to preparation of 50 or 100 returns every season. But once you have set your sights, try to conduct your practice and your dealings with clients in the same way as if you had already achieved the success you envision. For example, there is a certain confidence a successful professional exhibits, try to develop this confidence. There is a certain amount of joy and pleasure every successful professional derives from his work, try to convey this sense of joy to your clients as well. There is a genuine enthusiasm with which he or she tackles an unusual tax problem. Try to develop this enthusiasm within yourself, and let it be contagious, so that it influences everyone around you.

You will find that as soon as you have firmly and sincerely set your goal, it will influence your thinking from that moment on. For example, one of our more successful graduates years ago devised an effective tactic that still works wonders. It may be helpful to you too. In the first week of January he writes down every method he can think of to reach as many people as possible. He not only lists general areas such as co-workers, PTA, friends and relatives, but he works out practical methods of informing entire groups of prospective clients. For instance, he belongs to a men's club which has over 50 members. He makes an effort to talk to each of the members at least once or twice between January and March. Every year he also asks the president of the club that he be given a few minutes to address one of the meetings on a current tax concern. Ditto with gardening, church and various sports groups in which he is active. Then he tallies the total number of new clients he estimates he will be able to obtain through this method. Surprisingly—or perhaps not surprisingly—his estimates over the years have been remarkably accurate.

APPENDIX

Average Tax Return Fees

By Preparer Category

The average tax service fees below were abstracted from the survey report ***1998 Income & Fees of Accountants in Public Practice***, published by the **National Society of Accountants.**

The figures given are all-inclusive, covering both federal and state income tax returns, as well as the required supporting forms and schedules. Extra charges apply for returns requiring excessive time or additional research.

1040 Itemized

TP/PA (unlicensed)	$164
PA	190
EA	200
CPA	232
Overall Average	192

1040 Standard Deduction

TP/PA	$62
PA	83
EA	88
CPA	96
Average	86

1054 (Partnership)

TP/PA	305
PA	345
EA	357
CPA	410
Average	351

1020 (S Corporation)

TP/PA	408
PA	510
EA	484
CPA	519
Average	488

1020 (C Corporation)

TP/PA	427
PA	544
EA	506
CPA	566
Average	494

Hourly Fees for Tax Work

TP/PA	76
PA	84
EA	85
CPA	83
Average	82

Additional points:

- One-third of the practitioners surveyed offer electronic filing
- 77 percent charge an extra fee for representing clients at audits
- 54 percent base fees on hourly rates; 39% charge on fixed fee basis
- Average firm increased fees from '96 to '97 by 7%; Projected '98 increase was 6.5%
- Less than 10 percent accept credit cards

Copies of the 68-page Survey Report are available from the **National Society of Accountants** – 1-800-966-1574
Price: NSA members $45; Non-members $60

Suggested Tax Fee Schedule

The figures below are based on 1978 Tax Season fee schedules obtained from small to medium size accounting and tax firms. Where fee ranges are given, the smallest firms generally (but not always) tend to base their charges at the lower end. Some forms and schedules can vary greatly in amount of data and complexity and the fee will vary accordingly. Form 1040 figures include state tax preparation, where applicable.

At press time (October 1998) it is anticipated that 1999 Tax Season fees will rise by approximately five to eight percent.

Form or Schedule		Fee (or Range)
1040	U.S. Individual Income Tax Return – Standard Deduction	55-85
1040	U.S. Individual Income Tax Return -- Itemized	80-125
	Additional for: excess (more than 3-5) 1098s, 1099s,	2-3 each
	K-1s	20-35 each
	Alimony (received or paid); Pension-IRA disturb,	10-25 each
	Schedule C	50-100
	Schedule C-EZ	25-50
	Schedule D (1-6 items)	35-60
	Schedule D (additional items)	5-10 each
	Schedule D (additional for carryover computation)	15-25
	Schedule E (one rental property)	25-60
	Schedule E (additional rental properties)	25-40 each
	Schedule EIC	15-20
	Schedule F	45-100
	Schedule H	25-35
	Schedule R	10-20
	Schedule SE (short)	10-15
	Schedule SE (long)	15-25
1040A	U.S. Individual Income Tax Return (Short Form)	40-60
1040ES	Estimated Tax	25-40
1040NR	Nonresident Income Tax Return	100-275
1040X	Amended Income Tax Return (not including state return)	40-75
1045	Application for Tentative Refund (other than NOL)	35-50
1045	Application for Tentative Refund (Net Operating Loss)	50-100
1116	Foreign Tax Credit – Individuals, Fiduciary, etc.	40-75
1310	Statement of Person Claiming Refund Due a Deceased Taxpayer	15
2106	Employee Business Expenses	30-40
2119	Sale of a Home	30-40

2120	Multiple Support Declaration	5-10 each
2210	Underpayment of Estimated Tax	25-65
2441	Credit for Child and Dependent Care Expenses	30-40
2555	Foreign Earned Income	75-100
2555-EZ	" " "	35-50
2688	Application for Additional Extension to File	20-35
3800	General Business Credit	35-50
3903	Moving Expenses	10-25
4506	Request for Copy of Tax Form	10-15
4562	Depreciation and Amortization	40-125
4684	Casualties and Thefts	25-40
4797	Sales of Business Property (also involuntary conversions, etc.)	50-100
4835	Farm Rental Income and Expenses	25-50
4868	Application for Automatic Extension of Time to File U.S. Individual Income Tax Return	10-15
4952	Investment Interest Expense Deduction	60-75
4972	Tax on Lump-Sum Distributions	30-40
5329	Additional Taxes Attributable to Qualified Retrmnt Plans	25-40
6198	"At Risk" Limitations	40-60
6251	Alternative Minimum Tax – Individuals	60-85
6252	Installment Sale Income	40-65
8282	Donee Information Return	25-35
8283	Noncash Charitable Contributions	25-35
8332	Exemption Release – Divorced or Separated Parents	10
8379	Injured Spouse Claim and Allocation	35-75
8582	Passive Activity Loss Limitations	40-65
8582-CR	Passive Activity Credit Limitations	80-125
8606	Nondeductible IRA Contributions, IRA Basis, and Nontaxable IRA Distributions	20-35
8615	Computation of Tax for Children Under Age 14 Who Have Investment Income of More than $1,300	25-40
8801	Credit for Prior Year Minimum Tax	30-60
8814	Parent's Election to Report Child's Interest and Dividends	15-25
8815	Exclusion of Interest from Series EE Savings Bonds	20-35
8821	Tax Information Authorization	15-25
8824	Like-Kind Exchanges	40-65
8829	Expenses for Business Use of Your Home	40-60
9465	Installment Agreement Request	25-35

New Business
BRIEFS
PRACTICE BUILDING NEWS / IDEAS / STRATEGIES

♦ ♦ ♦ ♦ For Accountants & Tax Practitioners who want to attract new business, increase current billings, and upgrade their client base. ♦ ♦ ♦ ♦

Volume 2 Issue 3 March, 1997

Tax Practice is an extremely dynamic profession. Innovative practice building techniques, ideas and methods come up constantly, and so do opportunities for obtaining extra income from current clients, and offering new services that'll attract new clients.

Additionally, accountants and tax practitioners are rapidly shifting from pure bean counting to active business consulting, becoming involved in nearly all (non-technical) aspects of their clients' businesses or professions.

New Business Briefs, our popular monthly newsletter highlights the newest business building techniques and points out suggested areas and resources for consulting services. Bear in mind that you don't have to become an in-depth expert to make a sound contribution to your client's bottom line; what is more important is that you know where to get the specific information and advice. Combine that knowledge with your general business experience and savvy, and you can become a much sought after source of business (and personal, too) advice.

We are reproducing a few selected issues of the newsletter on these pages to round out the information presented in this volume. It is our hope that you'll be able to put it to most effective use.

Doing Business With The Government

Government - federal, state and local - provides a vast and often lucrative market for myriads of products and services. Yet, fear of the red tape involved or a belief that they are not big enough, has kept many small businesses and entrepreneurs from tapping this market. Yet, the fact is that an increasing number of governmental bodies are mandated by law to favor smaller entities and, as a result, have simplified the bidding process. Nevertheless, according to official statistics, thousands of contracts, worth billions of dollars, each year bypass small firms that do not know of or understand government buying and selling.

You can do your clients (and yourself) a great service if you familiarize yourself with the ABCs of selling to and dealing with the various governmental agencies. Bear in mind that if your client sells a marketable product or service, chances are that the government uses it. But it may require some digging to find out who, where and how.

The best way to quickly learn how the federal government procurement works is to visit the nearest office of the Small Business Administration (SBA). This agency is specifically charged with the responsibility to make sure that small business obtains a fair share of government contracts and subcontracts. For SBA locations call 1-800-8-ASK-SBA. But here is some basic information, with a few hints and suggestions to get you started.

How the Government Buys

Generally, to sell to the Government you must be on the solicitation mailing list for the product or service, or category of products or services you offer. These mailing lists are the basis for most federal civilian and military purchases. When soliciting for bids a civilian or military purchasing office sends bid invitations to the appropriate mailing list.

To get on the mailing list, each firm must show that it has the ability to fulfill contracts for the item or project. The bid invitations usually include all necessary specifications and instructions for the preparations of the bids. The contract is then awarded to the lowest qualified bidder. There are also provisions for *negotiated* contracts and for simplified procedures for purchases under $100,000.

How to Find Sales Opportunities

Two basic resources for information about government purchases are the *Commerce Business Daily* (available in many libraries) which lists proposed government purchases over $25,000, and the *US Government Purchasing and Sales Directory*. The directory lists what products and services the government buys, which agencies buy them, and whom to contact. It is available in some libraries, at all SBA offices, or may be purchased from the Government Printing Office.

The largest civilian purchasing agency is the General Services Administration (GSA). You can find out about items bought by the GSA by writing to, or visiting, one of the agency's Business Service Centers (call the SBA number above for locations).

Firms selling all type of business-related products or services by mail, via catalogs or other direct mail solicitations, have found government employees to be highly responsive to their offers. Most federal employees authorized to make purchases carry a special Visa-IMPAC credit card with a $2,500 per purchase spending limit and, according to latest figures, about 90 percent of these transactions are made over the phone. This, of course, simplifies matters considerably. Make sure that mailings targeted at the federal market carry the IMPAC logo (call 303-585-4793 for approval).
Again, the SBA can be very helpful in identifying prospective purchasers. Also consult a knowledgeable mailing list broker.

For selling at the state or local level, generally the best source of information and assistance is the nearest SBDA (Small Business Development Agency) office. Your State Commerce Department should also be able to help. Check the white pages of your phone book under "Government".

Instill a Marketing Culture

To help your practice grow - or even stay even - each staff member must make a conscious effort to promote your firm at every occasion. That doesn't mean that all partners and employees should drop whatever they're doing and start knocking on doors.

What is needed is the awareness that in the 90s pure technical skills are no longer enough to effectively compete,the ability to attract and keep business is equally important. Any staffer who internalizes this concept will automatically perform accordingly.

There are a number of ways you can help this along.

- Hire the right people. When interviewing prospective employees try to assess their interpersonal skills as well as their accounting or tax proficiency. Look for individuals who can communicate well and have leadership potential. Make it clear at the outset that their advancement in the firm depends heavily on their commitment and dedication to business building.
- Develop your staff members' marketing skills. Instruct them by word of mouth and by example how to build network and client relationships, how to ask for referrals, how to utilize their social contacts.
- Continuously keep emphasizing the need to expand business and revenues. During tax season and other high-pressure periods this part of the job is often forgotten and can easily remain buried.
- Provide ongoing encouragement and support. Acknowledge all successes and allow for failures. Criticize when necessary, but make it constructive - and private. Where appropriate, permit staff members to take key individuals to lunch or buy them gifts, tickets, etc.

Reward major successes swiftly. Give raises, bonuses, trips or other incentives to any employee who manages to snare an important new client or otherwise makes an significant contribution to the firm's profitability.

RESOURCES

Doing Business with the US Government a basic primer on the intricacies of selling to or contracting for Uncle Sam. Prima Publishing, 800-221-7945
The Truth About Money - a very helpful book when dealing with smaller, less sophisticated, clients because it addresses their concerns, such as investments, planning for retirement, long-term care, putting children through college, etc.. Explains the options, what to watch out for, advantages and disadvantages, etc. in plain English. $19.95, paperback. Georgetown University Press, 800-246-9606.
Valuing Accounting Practices - this book is a must if you are thinking of buying, selling, or merging. A bit on the heavy side, but extremely thorough and comprehensive. $85, hardcover. John Wiley & Sons. POB 6793, Somerset NJ 08875

NEW BUSINESS BRIEFS is published monthly by National Tax Publications
POB 382, Monsey, NY 10952
Phone 1-800-914-8138 • Fax 1-914-352-8138 • Website http://www.nattax.com
e-mail: ntts@mail.idt.net

New Business BRIEFS

PRACTICE BUILDING NEWS / IDEAS / STRATEGIES

♦ ♦ ♦ ♦ For Accountants & Tax Practitioners who want to attract new business, increase current billings, and upgrade their client base. ♦ ♦ ♦ ♦

Volume 2 Issue 9 September 1997

How To Increase Your Referral Business

A substantial amount of new business may come to you from *existing* clients in one of two ways: Your clients may provide you with new, additional work to do; or they may refer you to others. But another area worth pursuing for new clients is referral sources - consisting of those who are presently *not* clients of yours.

Don't overlook this Opportunity!

Business success today depends upon cooperation between a wide variety of people: partners, acquaintances, associates, and others you know. Interactions at various levels can be instrumental in building a strong business. Therefore, to develop a successful practice, take a little time out for the following:
Identify your most worthwhile and promising referral sources.
Cultivate relationships and associations with these resources. Make them a part of your life. In doing so, you may well discover marketing opportunities you never knew existed before.

It's a big world out there...

Obviously, it may be counter productive to just approach *anyone at all*. So *how do you* identify good referral sources?

For one, if you specialize in a particular area (or areas), then consider meeting others who share those interests and concerns. Attend functions, gatherings, and activities that focus on these special areas of interest. Mingle with those who are like-minded...at local or community-wide get togethers. Such an investment of time is a good first step.

If you have a specific topic to discuss, you may take the initiative to write or call a particular person in order to meet. This is a rather direct approach but depending on the personalities involved, it can be quite effective. You may begin by mentioning mutual friends or business associates and then continue from there.

To further broaden your base of referral sources, ask colleagues of yours to introduce you to others - at professional or social gatherings. Again, a certain amount of initiative is called for, but when you share common ground with others, you'd be surprised at how productive such encounters can be.

Organize your list of referrals.

Depending upon what factors you considered with regard to referral sources, that is how you should classify your referral names. For example, look at the following possible categories: neighborhood friends and associates; friends or acquaintances from school (i.e. those who shared a similar major or with whom you were members in a club); past teachers, former employees, employers, or co-workers; previous or ongoing business contacts. You only need to list a few names in each category. Then make the contact and take it from there. As you get referrals from those in your list, you may find your list and categories changing - with some names falling away while others are added. But this is how the procedure of referrals evolves.

This door swings both ways...

For a truly productive relationship with referral sources, make sure *you give them just as you get them*. When others see that you are actively referring customers, contacts, and clients *to them*... they will be much more likely to send referrals *your way* as well.

An extension of this is the idea of being ever personable, informative, and responsive to the situations of others. When you put forth the effort to share your professional expertise with others - when needed - it will be remembered and appreciated. Others will trust you and in return refer your services to others as well. Just as long as you are providing quality service and producing satisfied clients, you will reap the rewards of referrals.

Keep in touch...

The real payoff with referral sources comes as a result of meaningful and ongoing associations. If you can find and cultivate areas of common interest (i.e. a bowling league)...if you can get together at meetings or functions that express shared concerns (i.e. a mutual, charitable cause) ... if you can keep in touch with a newsletter or a holiday card... then the desired results will surely follow.

It's just like planting any seed: nurture it, expose it to the proper elements; in the end, it is bound to bear fruit.

Finding Your Niche!

You may not think of it in these terms, but non-profit organizations are a major economic force in the United States. At least, all statistical findings point in that direction.

What is a nonprofit organization? It's an organization that provides a service to society that is somehow being overlooked or neglected by for-profit organizations. This would encompass educational, religious, health, civic, and other social concerns. These organizations spend a substantial amount of money and employ an ever-growing workforce - all across the land. In fact, over one-fourth of all workers in the United States are employed by private and public nonprofit organizations. A power to contend with, wouldn't you say?

Helping those who help others

Nonprofit organizations tend to use generic accounting software to meet their financial needs, but the truth is they shouldn't. The specific rules and regulations governing such organizations really demand special personalized attention to handle their unique and often complex accounting needs.

As it turns out, the non-profit niche is an area that calls for a lot of financial attention. There are many types of accounting needs and services demanded: from reviewing fundraising materials and explaining how to read nonprofit financial statements ... to writing and reviewing grants to see if they meet government standards ...to developing budgets and participating in financial decisions.

Expertise and determination are called for...

The nonprofit niche is definitely worth considering, and although there is money to be made, you still have to proceed with caution. It's an area you have to become familiar with, and first-hand experience counts for a lot.

At first, that experience may come from volunteering for certain organizations. See, and get a feel for how the system works. Become known for what you do and then take it from there. In fact, the people you meet and work with may be prime referral sources to further your practice.

Once again, it's a way of branching out, and who knows, it just may turn out to be your particular market niche. In fact, many who work in the nonprofit domain report numerous benefits - besides the financial: the people, the causes, the environment, the inspiration and motivation they feel and experience...*also amounts to something!*

Worthwhile Tips To Keep In Mind...

- **Encourage marketing** ideas: You'll be surprised with what you'll come up with when you involve all your staff at your next marketing meeting. Brainstorming can produce wondrous results. Get everyone together, share ideas, encourage everyone to offer suggestions - even offer an incentive if a suggestion is actually used. Whether it's monthly or otherwise, try to adhere to a regular schedule of such meetings. The ideas that come out may be very useful, and as a morale booster, it can't be beat.
- **Increase response** to your standard Yellow Page ad by including a small photo of yourself. Research strongly indicates that this helps. We recently spoke to someone who reported that his picture was the most outstanding feature of his ad. People remembered the picture more than anything else, responded to it, and gained him instant recognition on a regular basis.
- **Get into the habit of** sending out short press releases to any neighborhood or community press that features such small articles. Be sure to mention any changes in your firm (be it one of location or expansion); make the public aware of any changes in law that may affect their finances; even try to develop a regular, informative column feature along these lines; and publicize any seminar or speech you may be giving or participating in. Address release to the Editor and provide a contact person with phone number in case there are any questions. Write short paragraphs, with most important info in first paragraph, with remaining material in order of importance (editors will often omit the last few paragraphs). Write concisely and to the point.
- **Do you have** a prospective-client interview coming up? Then do your homework! Prepare by finding out whatever helpful information you can - about your prospect, his or her line of work, etc. Impressing your prospective client with such insight and information will help you immensely.

Resources

The Under 40 Financial Planning Guide
Your younger clients demand a different financial-planning approach than the matureand senior group. This inexpensive 300+ page guide provides detailed, practical advice on paying off debt, buying a first home, building a savings plan and....yes, planning for retirement. $19.95 from Merrit Publishing 800-955-4775

WinWay Resumes for Windows
Whether you're trying to pitch your qualifications to a prospective client, advising a client on presenting a winning proposal, or attempting or attempting to help his fresh-out-of-college daughter find a job, this feature-filled software program can make your task much easier. Powerful, yet easy to use, it lets you custom tailor the same resume to the specific needs or requirements of the client, Abt. $40. WinWay Corp. 916-965-7878

If you're into assisting clients with obtaining financing - or planning to get into it (see "Getting Your Client the Loan He Wants...", Aug.'97 NBB), here are two very helpful software packages.

Loan Builder (Windows & Macintosh)
A well documented and simple to learn program, excellent for the novice in this line. With the help of the included JetForm Filler you can prepare most loan documents in no time. About $140. Available from JIAN 800-762-2413

Loan Watch (Windows)
If you don't mind a longer learning curve and don't need too much hand holding this powerful program has much to offer. It's especially helpful in analyzing what-if options for paying off loans early as well as in choosing between a variety of loan arrangements. About $150, from Compu-Share, Inc. 800-928-4256.

NEW BUSINESS BRIEFS
is published monthly by:
National Tax Publications
POB 382, Monsey, NY 10952
Phone 1-800-914-8138
Fax 1-914-352-8138
Website http://www.nattax.com
e-mail: info@nattax.com
Subscription price: $26.00 for one year; $48.00 for two years

New Business BRIEFS
PRACTICE BUILDING NEWS / IDEAS / STRATEGIES
For Accountants & Tax Practitioners who want to attract new business, increase current billings, and upgrade their client base.
VOLUME 3 ISSUE 3 MARCH 1998

HELPING TO PUT A ROOF OVER YOUR CLIENT'S HEAD...

Owning your own home where you can feel comfortable, secure, and where you are able to raise your own family – always has been and probably always will be – an essential part of the great American dream. But making this particular dream come true calls for a fair amount of financial savvy. There are a lot of factors to consider when making this long-term financial commitment and most homebuyers can benefit greatly from someone who can guide them through the intricacies of this particular financial maze, which can oftentimes be complex, confusing, and frustrating.

As a tax professional, you are in an excellent position to provide exactly this kind of assistance. Becoming familiar with the various ins and outs of mortgages so you can help your clients obtain one can generate a nice off-season income while it also helps strengthen your client relationships. Plus, consumers – such as your clients – need to know that mortgage brokers do not always have the consumer's best interests at heart. They oftentimes steer homebuyers to a particular loan, simply because they get a good commission from that lender.

In the role of tax professional or financial adviser, you may be the first to know that he is looking to finance or even refinance his home. You may then offer your services, in that you are familiar with, or have an expertise in this area and thereby offer your assistance. Of course, the more knowledge you have and the more connections you establish in the mortgage industry, the more you will have to offer your clients. Here are some resources to pursue that can help you develop your expertise in this area.

Taking advantage of today's computer technology is a great place to begin...

There is some very good inexpensive software as well as Web sites that can help you ...in turn help clients... in deciding the right type of mortgage to consider, what type of house he can really afford to buy, how many points, the amount of commission, and other costs to confront at closing. The Web is perfect for you if you're the kind of person who is online a lot. You'll find sites that will help you figure out every aspect of mortgages.

If you want to go back a step and get more familiar with the entire mortgage process, then visit The Mortgage Mart at www.mortgagemart.com. This site has sections loaded with descriptions of all available programs.

You can also help your client "prequalify" for a loan. Prequalifying means that a mortgage banker has looked over your credit history, income, and debts, and has determined the amount of credit for which you qualify. This makes your client more attractive to sellers. At www.dirs.com/mortgage/network, you are directed to a mortgage banker near you who can prequalify your client by phone, fax, or e-mail, within 24 hours. If your client has had credit problems, then jump to Loanicons at www.mindspring.com. to locate lenders who will help you obtain a new mortgage.

Once you obtain prequalification status you can look into the absolute lowest rate in the country by going to American Finance at www.financenter.com/homeloan/AFI/bestrate.htm. If you find a lower rate on the day you apply, the company will match that rate and pay you $100 on closing..

Let us now suppose that you understand the way mortgages work and you just want a way to figure out what your client will have to pay at different interest rates and for varying lengths of repayment. Then simply go to the Best Mortgage Calculators site at www.mortgage-net.com/calculators. After you plug in the numbers, you will be able to determine your principal and interest payments, compare the financial aspects of a 30 or 15 year fixed mortgage, and make other vital determinations as well.

Looking At Software...

If you are the type of professional who likes to have everything at your fingertips – when and as you need it – then take a look at some mortgage software. There are many inexpensive programs available that explain all aspects of obtaining, maintaining, or refinancing a residential loan.

Homebuyer – The Book and Software HomeBuying Kit (Stratosphere Publishing, 800-646-6456, $12) clearly explains the entire home buying process. In addition to doing all the number work for you, it provides definitions and explanations in an easy-to-understand English. The program also provides a state by state breakdown of mortgage related facts and figures. This program is thorough and extremely user friendly.

The Mortgage Kit (Dearborn Multimedia, 800-245-2665; Win 95, Win; $19.95 CD ROM, $39.95 CD and book) goes into great detail as it takes you through the steps of shopping for a new mortgage. It also explains very well all pertinent concepts and terms and takes you through various vital home buying scenarios.

Mortgage Plus Standard 2.5 (Bit Brain Software, 800-944-4248, Win 95, Win, Mac; $39) comes full of forms and worksheets, and calculators. This is a detailed, all-encompassing program with a wealth of worthwhile financial data.

And you didn't even have to leave your desk...

As a tax professional you know that there's a lot more to making all the proper calculations than meets the eye. Nevertheless, these Web sites and programs can help you immensely when it comes to helping your client work out a better deal more quickly, and all without leaving your desk.

Finally, this article dealt exclusively with residential mortgages. The commercial mortgage field is more complex, so we suggest you team up with someone professionally familiar with this area if you want to venture into it.

The Time Is Right To Consider - BUSINESS ARBITRATION!

Why business arbitration?

Picture one of the worst traffic jams imaginable. But now picture that traffic jam clogging up our courts rather than our city streets. That's right. With lawsuits ever on the increase and with overcrowded court dockets, small businesses very often have no real practical, realistic recourse when it comes to settling ongoing debt problems and disputes. Hiring attorneys is extremely costly and time consuming. It's stressful, eats away at the energies of both parties, and creates cash problems.

One manageable and cost-effective solution to all of this is third party intervention, or arbitration. Here is a workable solution that can open up lines of constructive communication between debtors and creditors. Moreover, arbitration becomes an extremely attractive option when it is offered to small business owners on a risk free, results oriented basis.

Arbitration actually allows the businesses involved to return their attention to where it belongs: to their revenues and profit generating activities. In most instances, the parties involved in a dispute do not know how to resolve a dispute. More often than not, it calls for the presence of an uninvolved third party who can create an atmosphere and establish the communication channels that can lead to satisfactory debt resolution.

This is where you, as a knowledgeable tax professional, can come in. Here is an area where you can put your financial expertise to productive use as you continue to develop your overall practice. The service you perform can benefit your client in any number of positive ways. With successful arbitration you put money back into the business immediately; you reduce your client's payables; and relieve the debtor from embarrassing and continuous collection calls. Your intervention can avoid an appearance in court, prevent the charges of legal fees, filing fees, and court costs, not to mention all of that precious time that is saved. And perhaps most important of all...your service enables your client to perform his necessary tasks – in order to stay in business and avoid possibly shutting down.

There are creditor benefits to consider as well. Your services can help return their costs of goods and services immediately in cash. Receive funds from businesses that would otherwise fold and not be able to pay. Raise instant working capital and improve their balance sheets. Your service preserves customers and retains profit from an otherwise lost customer / client. And oftentimes the creditor's business is saved.

Here is a genuinely profitable direction in which to develop your practice, and you only need to employ some basic negotiating skills that can be mastered in several ways. The following resource material should prove extremely helpful:

- **How To Mediate Your Dispute**: by Peter Lovenheim Paperback / published 1996.
- **Affordable Justice: How to Settle Any Dispute,** Including Divorce, Out of Court by Elizabeth L. Allen and Donald D. Mohr / Hardcover / Published 1997.

NEW BUSINESS BRIEFS is published monthly by *National Tax Publications*
POB 382 • Monsey, NY 10952 • Phone 1-800-914-8138 • Fax 1-914-352-8138
http://www.nattax.com • e-mail: nbb@nattax.com
Subscription price: $26.00 for one year; $48.00 for two years

New Business
BRIEFS
PRACTICE BUILDING NEWS / IDEAS / STRATEGIES
♦ ♦ ♦ ♦ For Accountants & Tax Practitioners who want to attract new business, increase current billings, and upgrade their client base. ♦ ♦ ♦ ♦

VOLUME 3 ISSUE 5 MAY 1998

HOW TO HELP CLIENTS REDUCE UNEMPLOYMENT RELATED COSTS

If you're like most people, you find yourself looking for ways to save money - however, wherever, and whenever you can. This applies to your private life, your professional life, and it should hold true as it relates to your clients' finances as well. One way to save money for a client and his or her company is to reduce tax liabilities. For example, there are numerous ways to control and cut unemployment tax expenditures, which this article will go on to discuss.

The fact remains that most payroll or unemployment taxes are not controllable. Some are fixed rate taxes such as the employer's portion of Medicare and Social Security taxes. However, unemployment tax is controllable since the rates are based on a company's experience. This is where steps can be taken to reduce a client's expenses. By implementing a few practical strategies, you can help your client reduce these costs. Here's how...

Get a hold of the facts.

Employers generally think that if an employee quits or is fired for good reason, that he or she is not eligible for unemployment benefits, but this is not so. Every state has its own criteria for unemployment eligibility; so learning what policies govern the state you are dealing with is the first step to consider in developing a business plan. Being aware of the policies in your state will enable you to act on your client's behalf. Whether this means properly processing claims for part-time workers, or applying for relief from charges...the end will be a reduction in unemployment costs.

Savings in auditing reports...

You can also come across savings in the following way. When your client receives benefit charge statements from the state, the statements should be audited to determine the following: employee names and social security numbers should be correct. Be sure the employees on the report are eligible for benefits. Make sure that claimants have not exceeded their maximum benefit allowances. Make sure benefit charges are correct. Be sure credits are reported accurately.

Unfortunately, state agencies tend to make mistakes and your client pays the price if the errors are not caught.

Be in attendance at hearings.

It may seem cumbersome, even time consuming, but representing your client at unemployment hearings can also be financially rewarding. Very often it is at hearings that eligibility determinations are reversed. The more aware you are of unemployment laws, the more you are in a position to influence what takes place at hearings. Former employees may receive benefits simply because your client wasn't prepared with the proper documentation, and this is where you can be of help.

Rehire ex-employees.

One effective way to reduce unemployment expenses is to offer jobs to former employees. A former employee's refusal to accept such an offer can result in a suspension of benefits. An acceptance of the offer often means that the employer does not have to restart the taxable wage base. Either way, the result is a savings for the employer. Of course, this only works if your client really wishes to rehire the employee in question.

Knowledge, rightfully employed, is a powerful thing. You may not be able to control all of your client's payroll taxes, but there are some things you can do to control your unemployment taxes. Through a set of strategies that include the suggestions mentioned above, you can ccrtainly help clients reduce their costs for unemployment taxes.

SELLING: IT'S A PART OF LIFE...

Accountants and tax practitioners often make this distinction: they don't mind the "marketing" part of their business but don't care much for the "selling." What's the difference? And as a tax professional, do you know when to market and when to sell?

One convenient way of looking at the difference is this: Marketing involves formal promotional efforts that primarily make use of mass media... be it radio, television, newspapers, magazines, brochures, the internet, etc. In one or more of these ways you are letting people know what you do and what services you provide.

Selling is a personal process whereby you try to persuade someone to engage in the services you offer. You are trying to get a prospect to transfer money over to you in exchange for the professional expertise you can offer him. Not for all, but for many, this becomes an awkward, even difficult and uncomfortable part of the profession. But in spite of the fears and dislikes, selling skills continue to be an important part of a tax professional's arsenal of tools and devices needed to gain and maintain a growing clientele.

There are numerous reasons to employ selling skills. For one, the competition is getting intense and clients demand value. They want to know precisely what you have done for them. What gain was achieved by your providing a specific service?

As a tax professional, you have to be ever on the lookout to provide the right types of client services. If you check the ages of your client base and look for averages you may find that you should be marketing and selling specific services such as retirement planning. Clients today continually want more and different services, and you have to be ready to respond...or someone else will. And if you can provide certain services, then you have to let your clients know about them.

Many tax professionals simply may not be prepared to think in these terms. Since when are "selling skills" needed to do effective tax work? "It's not what I was trained for," you may be saying. Consequently, many don't know how to sell, and simply assume that clients will always be calling on them. So even before looking into selling skills, there has to be a shift in focus and recognition that this is an essential step toward professional growth.

In the meantime, here are some areas where selling skills come in handy.

- In the client relationship, be a good listener. A client, especially new ones, may need to let off some steam, or explain something in detail. Your ability to simply listen and commiserate can serve as an excellent personal selling point.
- Always be civil, respectable, and straightforward with clients. Explain clearly each of your fees. If there is a fee increase, explain what it is, and what increase in value the client will gain thereby. If any problem or concern arises, don't procrastinate. Show the client that you can deal with it immediately. Unresponsiveness leads to discontent, while your immediate response leads to contentment.
- Don't take your clients for granted. An occasional call or periodic card lets them know that you are thinking of them. When a client first came your way you may have serviced him with enthusiasm, and then with time, that excitement faded. You might say this is the nature of any relationship. However, like all relationships, that excitement can be rekindled. You can continue to provide that kind of interest and attention – and it doesn't call for a great expenditure of effort. An occasional note, card, or call – to mark a special event or holiday – is enough to show that you care enough to make a connection.
- When it comes to your professional relationship with clients, try to include them in any decisions or determinations you are making. Even if you already know the answer to a certain question, contact them to double check or see if any data has changed. This shows that you are actively working on and concerned about their financial situation. If you have new information that may be beneficial to a client, be sure to share it with them – by card, letter, call, or e-mail. This too shows that you have the best interests of your clients in mind.
- You may want to send out a brief, periodic letter of updates including information that may be useful to all clients. In such a letter you may also ask for feedback with regard to your services. This shows that you are concerned with the quality of your work. And when a client responds with a critique, acknowledge it personally and thank him for the response.
- If appropriate, share personal experiences that may bolster your relationship with clients. Offering your own experiences or drawing upon your own background and bringing it to bear on the business relationship may add a personal dimension to your overall relationship. You have to be careful in this regard, but there are certain advantages to be gained when you expand your relationship beyond the strictly technical to make it more personal.

All of the above can serve as selling points that can either bring in new clients or enhance present client relationships. If you employ them wisely, they can certainly serve to better your business.

NEW BUSINESS BRIEFS, is published monthly by:
National Tax Publications • POB 382 • Monsey, NY 10952
Phone 1-800-914-8138 • Fax 1-914-352-8138
Website http://www.nattax.com • e-mail: nbb@nattax.com
Subscription price: $26.00 for one year; $48.00 for two years

TABLE OF CONTENTS